AF809431

Contemporary Issues in Science Communication

Series editor: **Clare Wilkinson**, University of the West of England, Bristol

---

As science communication continues to establish itself as a discipline in the 21st century, this book series links the present with the past to develop our understanding of its history, both in practice and as an academic discipline.

Out now in the series:

Queering Science Communication
Edited by **Lindy A. Orthia** and **Tara Roberson**

Race and Sociocultural Inclusion in Science Communication
Edited by **Elizabeth Rasekoala**

Find out more at:

bristoluniversitypress.co.uk/contemporary-issues-in-science-communication

International advisory board:

Alison Anderson, University of Plymouth, UK

Ayman Elsayed, Grand Egyptian Museum, Egypt

Birte Fähnrich, Berlin-Brandenburg Academy of Sciences and Humanities, Germany

Carla Almeida, Museu da Vida, Fundação Oswaldo Cruz, Rio de Janeiro, Brazil

Daria Denisova, Information Technologies, Mechanics and Optics University, Russia

Elizabeth Rasekoala, African Gong, The Pan African Network for the Popularization of Science and Technology and Science Communication, South Africa

Emily Dawson, University College London, UK

Erik Stengler, The State University of New York College at Oneonta, US

Jenny Martin, University of Melbourne, Australia

Lesley Markham, Center for Open Science, US

Marina Joubert, University of Stellenbosch, South Africa

Padraig Murphy, Dublin City University, Ireland

Find out more at:

bristoluniversitypress.co.uk/
contemporary-issues-in-science-communication

CONTEMPORARY ISSUES IN SCIENCE COMMUNICATION

SCIENCE COMMUNICATION AS SUSTAINABILITY

Transformative Practices and Arts-Based Approaches

Edited by
Sabrina Vitting-Seerup, Martin Grünfeld, and Marianne Achiam

With a Foreword by
Sarah Davies

BRISTOL
UNIVERSITY
PRESS

First published in Great Britain in 2026 by

Bristol University Press
University of Bristol
1–9 Old Park Hill
Bristol
BS2 8BB
UK
t: +44 (0)117 374 6645
e: bup-info@bristol.ac.uk

Details of international sales and distribution partners are available at bristoluniversitypress.co.uk

© Bristol University Press 2026

DOI: 10.51952/9781529247268

British Library Cataloguing in Publication Data
A catalogue record for this book is available from the British Library

ISBN 978-1-5292-4724-4 hardcover
ISBN 978-1-5292-4725-1 ePub
ISBN 978-1-5292-4726-8 ePdf

The right of Sabrina Vitting-Seerup, Martin Grünfeld, and Marianne Achiam to be identified as editors of this work has been asserted by them in accordance with the Copyright, Designs and Patents Act 1988.

All rights reserved: no part of this publication may be reproduced, stored in a retrieval system, used to train or develop an artificial intelligence system, or transmitted in any form or by any means, electronic, mechanical, photocopying, recording, or otherwise without the prior permission of Bristol University Press.

Every reasonable effort has been made to obtain permission to reproduce copyrighted material. If, however, anyone knows of an oversight, please contact the publisher.

The statements and opinions contained within this publication are solely those of the editors and contributors and not of the University of Bristol or Bristol University Press. The University of Bristol and Bristol University Press disclaim responsibility for any injury to persons or property resulting from any material published in this publication.

Bristol University Press works to counter discrimination on grounds of gender, race, disability, age and sexuality.

Cover design: Liam Roberts Design
Front cover image: Sofie Louise Dam
Bristol University Press uses environmentally responsible print partners.
Printed and bound in Great Britain by CPI Group (UK) Ltd, Croydon, CR0 4YY

FSC
www.fsc.org
MIX
Paper | Supporting responsible forestry
FSC® C013604

Bristol University Press' authorised representative in the European Union is:
Easy Access System Europe, Mustamäe tee 50, 10621 Tallinn, Estonia,
Email: gpsr.requests@easproject.com

Dedicated to everyone fighting for a more
just and liveable world

– and to the hope that still grows in
unexpected places

Contents

Series Editor's Preface

Clare Wilkinson

As science communication continues to establish itself as a discipline in the 21st century, there has never been a better time to consider contemporary science communication and its practices. Some years on from the establishment of this series, contemporary scientific and technological developments continue to raise significant new questions for communicators, including quandaries surrounding how to best address controversies associated with science, as well as the broader social and ethical challenges they often involve. Meanwhile, topics like mis and dis-information, trust, and the role of social media appear to be rising in importance in current social, cultural, and political contexts, with technologies like GenAI surpassing models we may have relied on in the past or any sense of academic resolution. Equity, diversity, and inclusion have been reinforced in terms of their importance, when political infighting and global events increasingly appear to destabilize their foundations.

'Contemporary Issues in Science Communication' is a multidisciplinary series, welcoming submissions from a wide range of disciplinary areas, and international in outlook. Books contained in the series seek to be engaging, straightforward, and conversational in style, and of interest to science communication practitioners as well as academic audiences.

Books published in the 'Contemporary Issues in Science Communication' series consider such science matters and their relationships to communication, engagement, and broader social conversations. The series links the present with the past by publishing titles that develop our understanding of the history of science communication, both in practice and as an academic discipline. However, they also cover a range of topics relevant to contemporary science communication, including, but not limited to: definitions, history, and ethics of science communication; expertise, replication, and trust; interdisciplinary knowledge; ideologies; knowledge and new forms of media; public policy; gaming, sci-art, and visual communication and inclusivity in science communication. The scope of the series is broad but so are the challenges facing science communicators and public engagement practitioners.

In this book, *Science Communication as Sustainability: Transformative Practices and Arts-Based Approaches*, edited by Sabrina Vitting-Seerup, Martin Grünfeld, and Marianne Achiam, we find a collection which directly tackles one such challenge: that of the role of communication in sustainability, an ongoing topic of relevance to both science communicators and us all in the context of the climate crisis. In tackling this subject, the authors offer a highly novel approach, encouraging readers to consider new terminology and frameworks to transform the role of science communication in sustainability, as well as a suite of methods that tackle this topical and important subject in creative and innovative ways. Drawing together an international range of authors, coming to the subject from different disciplinary as well as academic and practical perspectives, the book explores a range of creative examples while proposing actions for science communication as a discipline and community. Since the collection starts with an engaging invitation from Sarah Davies to explore the book further, I will leave the specifics of my introduction there.

I hope you will find this book intriguing and hopeful but also realistic about some of the significant collective challenges we face in relation to sustainability. If you are interested in becoming involved in the series personally, please do get in touch. Finally, thank you to Sabrina, Martin, and Marianne, as well as all their contributors, for their enthusiasm to be a part of this series and for contributing such a creative and important collection.

List of Figures and Tables

Figures

Table

Notes on Contributors

Marianne Achiam (Associate Professor) leads a research group on science communication, and her work constitutes the starting point for the research programme 'Addressing Sustainability with Arts-Based Science Communication'. Her research and teaching focus on using interdisciplinary methods to create effective science communication for sustainability. She has published research in international, peer-reviewed journals in the fields of museology, science education, and science communication, has authored a number of chapters for peer-reviewed anthologies, and has co-edited books published by Springer and Routledge.

Maria Brænder is a transdisciplinary artist based in Copenhagen who works in the cross-currents of sound/performance/installation (bio)art exploring and manifesting the ideas of radical environmental philosophy through installation (and/or performance) experiences for audiences. Currently a PhD candidate at the Royal Conservatoire of Scotland, Brænder is undertaking practice-based research in how lived experiences of multi-being relationality (informed by science, the sensuous, and the affective) can reverberate into art making.

Sofie Louise Dam is a comic artist and illustrator based in Copenhagen. She has a background in classical drawing and a BA in Graphic Storytelling from The Animation Workshop. Her work also spans children's books and comics, but she has a passion for nonfiction storytelling and visual co-creation. She has contributed to 'The world is in you' art-science exhibition at Kunsthal Charlottenborg and has previously collaborated with researchers on projects that explore sustainability through arts-based approaches.

Adam Dickinson is a poet and professor whose work probes the porous boundaries between science and poetry. Through experimental practices, such as chemical and microbial analyses of his own body, his writing explores how environmental systems, personal health, and language intertwine. His award-winning collection *Anatomic* exemplifies his commitment to research-driven poetics that question how we metabolize the world. He teaches creative writing and poetics at Brock University in Ontario, and his work has been featured in international festivals and translated into several languages.

Martin Grünfeld has a PhD in philosophy from University College Dublin but often finds himself working in the intersections between disciplines and engaging in transdisciplinary collaborations with artists, curators, and other organisms. For the past five years, he has worked as a curator and researcher at Medical Museion. Currently, he is developing his research in arts-based science communication focusing on issues around climate change and sustainability through sound, art, and poetics.

Jacob Thorek Jensen (PhD Fellow) has considerable experience in science communication through his work as a museum curator since 2015. His practice focuses on inclusion in cultural institutions and the museum's potential as a site for co-creation and public engagement. Drawing on his background in curatorial work, his PhD research explores how science museums can develop spaces and practices that support collective imagination around sustainable futures. With a strong understanding of both institutional structures and visitor engagement, he bridges theory and practice to rethink the role of science museums in times of social and environmental change.

Alison Pouliot is an ecologist, writer, and photographer who brings hidden worlds into focus – particularly the fungal, the spineless, and the overlooked. Working between hemispheres and habitats, her interdisciplinary practice combines scientific enquiry with evocative storytelling to foster new ways of seeing and valuing life. Through immersive fieldwork and visual narratives, she explores the textures, moods, and forgotten corners of ecological systems, advocating for their care and conservation with both precision and wonder.

Sabrina Vitting-Seerup (Assistant Professor) explores the intersections of science, culture, and communication. Her research, grounded in Cultural Studies, examines how arts-based methods, emotional engagement, and sensory experiences shape our engagement with science. She currently focuses on transforming sustainability science communication through creative, interdisciplinary approaches. Alongside her academic career, she has led a science communication consultancy for nearly a decade, working with a range of researchers, organizations, and audiences to make complex research accessible, resonant, and impactful.

Louise Whiteley (Associate Professor) is interested in how ideas about ourselves and our place in the world get formed between biomedical science and everyday cultures. She makes collaborative exhibitions and events alongside her research at the Medical Museion in Copenhagen, where she has a focus on how we communicate ethos and emotions when we talk about science.

Acknowledgements

This project was generously funded by the Novo Nordisk Foundation under grant number NNF22OC0078707 as part of the research programme 'Addressing Sustainability with Arts-Based Science Communication'. We are grateful for their support in enabling interdisciplinary collaboration between artists and researchers working to reimagine how sustainability is communicated and understood.

Foreword

Sarah Davies

It is a truism, at this point, to say that science communication is an important part of contemporary societies. At a moment in which the lives of many are being impacted by environmental breakdown and climate crisis, when technology is driven by oligarchy, and when we must all make choices about how to respond to expert advice on everything from health to education, it is clear that technoscience imbricates multiple aspects of our lives. We must make individual and collective decisions about how to navigate these environmental, technical, and scientific developments, and it is the role of science communicators to help us do this.

As this volume so powerfully shows, one vital area for science communication that takes on this task is that of sustainability – efforts to ensure that human and non-human life is sustain-able, perhaps even having the possibility of flourishing. In taking sustainability as its focus, the collection responds to one of the central challenges of our times – quite simply, how we might collectively find ways of surviving, of halting or mitigating the extinctions, extractions, disasters, and exploitations that are being experienced around the world. As the editors of this book write, what sustainability fundamentally calls for is a reimagining of how to be in this world. The chapters that follow offer a rich set of suggestions and provocations about how this might be done, challenging standard assumptions in both sustainability and science communication that it is only human life that matters, that scientific knowledge is the sole reliable way to navigate the crises that we find ourselves in, or that emotion or art-based practices should be sidelined in favour of rationality. In foregrounding topics and practices such as artistic methods, emotion, imagination, Indigenous futurism, resonance, and relationality, the chapters rethink science communication's theory and practice, both in the urgent context of sustainability and more widely.

The collection therefore adds not only to discussions of best practice in environmental communication but to a body of recent scholarship that has offered fresh theorizations of the nature of science communication. Work that has framed science communication as culture, conversation, event, or (eco)system (for example, Horst and Michael, 2011; Braun and Könninger, 2017; Davies et al, 2019; Bucchi and Trench, 2021) has, in different ways, sought to move the field on from discussions of the 'science of science communication' or of contrasting 'deficit' and 'dialogue' models. In their varying ways, each of these approaches seeks to recontextualize science communication activities, presenting them not only as concerned with the flow or creation of scientific knowledge but as one aspect of wider societal dynamics. This volume furthers these discussions in two important ways: on the one hand, a view of science communication not of or about sustainability but *as* sustainability highlights public communication's performative dimensions. Sustainability cannot only be the topic of communication activities; it must be enacted by them: '[W]ith our proposition to reimagine science communication as sustainability', the editors write, 'we seek to make sustainability happen'. On the other, using the metaphor of terroir to think science communication (as outlined in Chapter 2) foregrounds relationality and the embedding of any science communication practice not only in specific social and cultural contexts but in webs of non-human (perhaps even non-living) actants and in particular spaces and places. Terroir helps us to see science communication as a heterogeneous assemblage in which many different actants participate, and which is always contingent on the specificities of particular sites and temporalities. It thus contextualizes science communication even further, allowing us to see (and study, and practice) it not as a universalized phenomenon but as a local and relational one.

The chapters that follow expand on these ideas and add many others – using history or sound to rethink science communication, for instance, or highlighting the generative (in contrast to appropriative) potential of engaging with artistic practices. If I might add one further node to this mesh of interconnected ideas, it would be that of care – an idea and practice around which many of the contributions circle. If care is 'a species activity that includes everything that we do to maintain, continue, and repair our "world" so that we can live in it as well as possible' (Tronto, 1998, p 16), then the view of science communication that is presented here is certainly one that is care-ful. Our world cries out for care right now, as humans, non-humans, and the material environments that sustain us all suffer in myriad and profound ways. This volume offers important suggestions as to how science communicators might respond to these needs and develop a practice of care-ful, responsive communication that enacts a more liveable world.

References

Braun, K., and Könninger, S. (2017) 'From experiments to ecosystems? Reviewing public participation, scientific governance and the systemic turn', *Public Understanding of Science*, 27(6): 674–89.

Bucchi, M., and Trench, B. (2021) 'Rethinking science communication as the social conversation around science', *Journal of Science Communication*, 20(3): Y01.

Davies, S.R., Halpern, M., Horst, M., Kirby, D.A., and Lewenstein, B. (2019) 'Science stories as culture: Experience, identity, narrative and emotion in public communication of science', *Journal of Science Communication*, 18(5): 17.

Horst, M., and Michael, M. (2011) 'On the shoulders of idiots: Re-thinking science communication as "event"', *Science as Culture*, 20(3): 283–306.

Tronto, J. (1998) 'An ethic of care', *Generations: Journal of the American Society on Aging*, 22(3): 15–20.

1

We Need This Book: An Introduction

Marianne Achiam, Martin Grünfeld, and Sabrina Vitting-Seerup

We live in an age where human activities have altered the global environment to the extent that the life-critical zone of the Earth is beginning to break down. These activities – first and foremost the burning of fossil fuels, but also the over-exploitation of natural resources through the acceleration of technology, industry, and agriculture (McNeill and Engelke, 2014) – are rooted in the ensemble of socio-cultural norms, discourses, attitudes, and practices that characterize the modern era (Bonneuil and Fressoz, 2017). Even though modernist ideals, such as individualism, capitalism, industrialism, and scientific and technological progress, have undisputedly improved the lives of millions of people, they have also had a range of unintended consequences that are now threatening humanity. Today, these ideals remain firmly embedded in Western ways of thinking and doing; this means that changing them requires principled attention and radical action. This book is our contribution to focusing such attention and prompting such action. Specifically, we argue that science communication has a fundamental role to play in changing our current, shared trajectory in the direction of a sustainable future.

To begin to address problems such as anthropogenic climate change, the biodiversity crisis, and the looming threat of new zoonotic pandemics, it immediately seems clear that the standard model of science won't do. Its dominant frameworks have largely treated the natural world as an external object to be controlled rather than as an entangled system that we, as humans, are part of. Indeed, scholars have critiqued the modernist vision of science, arguing that its rootedness in objectivity, reductionism, and technological progress has directly contributed to the crises we now face (Dürr et al, 2005; Beck, 2016; Stengers and Muecke, 2018). As a result, new models of critical, post-normal sustainability science have emerged, calling for more participatory, contextual, and ethically engaged approaches to knowledge production (Funtowicz and Ravetz, 1993). At their core, these models challenge the boundaries of what counts as science and who gets to participate in scientific knowledge making.

This shift has profound implications for science communication. If science itself must become more participatory, then its communication cannot remain a unidirectional transmission of facts. Science communication, understood as the social conversation around science (Bucchi and Trench, 2021), has long operated on the assumption that scientific knowledge is neutral, rational, and self-evident. Yet history shows that even when scientists have warned of global crises (Fourier, 1837; Arrhenius and Holden, 1897), their insights have failed to curb unsustainable behaviours. This is, at least in part, because mainstream science communication assumes that presenting 'the facts' should be enough (Nielsen, 2013; Takach, 2016; Canon et al, 2022). However, as Hochschild (1979) has convincingly shown, decision-making is shaped by emotions, lived experience, social norms, and cognitive biases. If sustainability science is to succeed, its communication must move beyond an information-deficit model and towards an approach that engages diverse ways of knowing to foster collective problem solving.

When we point to a new role for science communication to support the democratization of sustainability science, we are calling for a model of science communication that can encompass and incorporate a wide range of experiences and ways of knowing and engage a variety of interested and/or affected parties in the making of knowledge. In the following, we will propose a new avenue for science communication reimagined *as* sustainability. However, before we begin our conceptual work, we define the contested term 'sustainability' that we have built this book around.

Defining sustainability

In seeking to make sustainability more than just a topic for dissemination and dialogue, we must first consider what the concept means and how it potentially shapes science communication. 'Sustainability' is a cosmopolitical idea travelling across diverse fields of research, political and economic systems, media and marketing, businesses and agriculture. Indeed, it has become one of the defining yet also ambiguous buzzwords of our time.

Going back to its inception in the Brundtland Report (1987), sustainable development was framed as 'development that meets the needs of the present without compromising the ability of future generations to meet their own needs' (p 41). Here, sustainable development is a normative principle that prescribes ways of living and governing that do not endanger the well-being of future generations (Godemann and Michelsen, 2011). 'Sustain-able', in this sense, implies that the human population will be able to maintain similar living conditions and economies in the future. However, while sustainable development focuses on the political and economic challenges of improving socio-economic conditions, particularly in underprivileged regions, sustainability as a broader concept has wider implications, shaping what Loura and Dickinson (2016) call a 21st-century mindset.

For Loura and Dickinson (2016), sustainability concerns a shared commitment to managing and caring for the limited resources we depend on, the aspirations and rights of the human population, and continued economic growth. Although far-reaching, this definition also reveals its own limitations. By framing sustainability primarily through the lens of development and resource management, we risk reinforcing a disembodied and anthropocentric view of the world, one in which human needs remain central and the environment is treated as an externality to be optimized (compare Bonneuil and Fressoz, 2017) rather than as an interconnected system. Let us be clear: this is not the sustainability we name on the cover of this book. This tension found within the origin of the term points to a deeper problem: when sustainability is primarily articulated in the language of management, optimization, and growth, it risks becoming a flexible label that can be attached to practices and policies without fundamentally challenging the worldviews that produced the current crises. In such framings, sustainability remains compatible with extractive logics and techno-managerial solutions, even as they continue to erode ecological relations and multispecies livelihoods. What gets lost in this translation is an understanding of sustainability as a question of how humans relate to the more-than-human world – not as managers of it, but as entangled participants within it.

A more general point that follows from this is that sustainability becomes an empty signifier when it is detached from ecological thinking and our relationship to the more-than-human world. Dominant interpretations of sustainability tend to be highly technical, emphasizing better management of the non-human world rather than a fundamental shift in how we relate to it (Van der Ryn and Cowan, 2013). However, sustainability is not only about preserving resources for human use – it is about rethinking our place within ecological systems. The European Commission (2022) offers a definition that reflects this shift: '[S]ustainability means prioritizing the needs of all life forms and of the planet by ensuring that human activity does not exceed planetary boundaries.' This formulation moves beyond an anthropocentric model and instead acknowledges the relational, interdependent nature of all life. It aligns with what Alexander and others have called 'deep sustainability' – an approach that not only addresses systemic issues but also questions the underlying cultural, philosophical, and epistemic assumptions that structure human–environment relations (quoted in Roös, 2021).

The notion of *deep sustainability* makes clear that sustainability is not a matter of incremental adjustment but of profound reorientation: a shift towards recognizing human–more-than-human entanglements, the limits of extractivist worldviews, and the necessity of relational modes of living. Science communication, we argue, requires a transformation of the same depth that reorients not only what we communicate but how, with whom, and to what end.

Science communication as sustainability

Our titular call to rethink science communication *as* sustainability arises both from the scale and systemic nature of the crises we face, and from the limitations of existing approaches within sustainability communication scholarship to grapple with their epistemic, political, and relational dimensions. Within the literatures on sustainability communication, there is a morass of prepositions that shape how sustainability communication takes place either as a mono-directional transmission *of* sustainability issues or as a horizontal dialogical exchange *about* sustainability (Godemann and Michelsen, 2011; Fischer et al, 2016). However, communicating sustainability issues can also (and most often does) involve a normative dimension – communicating *for* sustainability, where the preposition *for* indicates that such forms of communication are explicitly and directly advocating sustainability and designed to facilitate sustainable development (Fischer et al, 2016). But while communication *for* sustainability is often described as drawing on existing models of sustainability communication (be they mono-directional or dialogical), our contention is that it is not enough to draw on the pre-existing communication models.

Building on Lyle (1996), we argue that sustainability is not just another subject that can be packaged nicely and communicated within dominant science communication models such as the deficit and dialogue models. Rather, it calls for a fundamental change in the way we view science communication: we must attempt to cultivate a culture of sustainability (Mochizuki and Yarime, 2016) to make a difference and push the sustainability agenda. Or more radically formulated, we need to challenge the fundamental ways in which we think about science communication, and that is our main task in this book. This claim runs parallel with Sterling's systems approach to learning *as* sustainability. For Sterling (2004), sustainability is not just an add-on to existing educational structures and curricula but implies a fundamental change in educational thinking and practice – it requires transformative learning. In contrast to the *of*, *about*, and *for*, which adds ideas of sustainability within existing frameworks, to reimagine science communication *as* sustainability entails a transformation. Yet like Sterling's nested view of the learning process, we do not want to rule out other forms of sustainability communication; rather, we propose to think of them as nested within yet radically reshaped by our notion of science communication as sustainability. In that sense, science communication *as* sustainability involves a double move: it takes sustainability as its main topic (it is *about* sustainability), and it is shaped by what it communicates (it is enacted *as* a form of sustainability).

If we for a moment step back into the media theory of McLuhan, perhaps we can make sense of this apparently weird proposition by way of his famous dictum 'the medium is the massage'. For McLuhan, the medium – be it handwriting, electrical technology, or social media – works us over (McLuhan and Fiore, 1967). What he means by this provocative phrasing is that our thoughts and actions as well as societies are shaped by the media we use. Indeed, media extend and shape our perception of the world. This entails that media are not neutral and innocent vehicles for information but rather shape it and, along with it, our relationship with the world. Likewise, we could argue that the communication models we use are not innocent but shape not only how ideas of sustainability travel but also the very ideas of sustainability. So, our proposition to think science communication as shaped by or embodying sustainability entails a move from thinking about the communication model as neutral and passive. It calls for us to take seriously the fact that the model becomes (part of) the message as it is massaged into the bodies and minds of publics. In that respect, science communication as sustainability involves a move from representation to performativity: science communication should not just represent sustainability but enact it.

Here it is crucial to distinguish between *performance* and *performativity*. While performance has emerged alongside other art–science combinations within science communication in recent years (Bucchi and Trench, 2021), such approaches remain niches or additions to the science communication toolbox. Meanwhile, our suggestion is more wide ranging – it entails a reshaping of science communication itself towards a paradigm of performativity with its own knowledge logics and positionality of publics and communicators. To think of science communication in terms of performativity means reframing it as not just communicating knowledge contents but also as doing something like Austin's (1963) performatives. Even though Austin's notion of the performative was originally devoted to a category of special utterances such as 'you are under arrest', the concept of performativity travelled across the humanities and in that process became generalized to describe all kinds of cultural phenomena and events that happen because someone does something recurrently (Bal, 2009).

When we propose to rethink science communication as sustainability, we are trying to turn sustainability into more than just a topic, namely a recurrent event that transforms the way we perceive the world and our place in it. This proposition aligns with Horst and Michael's reframing of science communication as an *event* – as a coming together of different entities, be they social and material, human and non-human, micro and macro, cognitive and affective (Horst and Michael, 2011). What this view entails is that science communication itself is produced in a mixing of relations that constitute it as science communication – it is emergent and performative. Building on this, perhaps we could say that with our proposition to reimagine science communication as sustainability, we seek to make sustainability happen.

Cultivating a different mindset

Yet it is easy for our thinking and imagination to get caught in the beaten track (Heidegger, 1968). How, then, do we seek out alternative pathways through the wicked landscape of sustainability communication? When we suggest reimagining science communication *as* sustainability, our proposition itself functions as a simile that connects two seemingly distinct concepts, enabling us to view science communication through the lens of sustainability and vice versa (compare Margolis, 1957). This is more than a rhetorical gesture, however. Science communication and sustainability are interconnected practices, actively shaping and transforming each other. Just as sustainability science conceptualizes socio-environmental issues as coupled human–environment systems (Clark, 2007), science communication should embrace the understanding that communicators, audiences, and researchers are embedded within complex sustainability challenges. Our role is not to transmit information but to actively participate in shaping knowledge, relationships, and futures.

To reimagine science communication as sustainability calls for a transformation of how we know, relate, and act in the world and simultaneously call into question the adequacy of the dominant communicative forms, media, and methods through which science communication has traditionally been enacted. Models oriented towards information transmission, rational deliberation, and linear persuasion are poorly equipped to engage with the affective, sensory, ethical, and relational dimensions of sustainability challenges. If science communication is to enact sustainability rather than just describe it, it requires forms that can cultivate complexity, uncertainty, emotional investment, and more-than-human relations.

In this book, we focus specifically on arts-based methods not to suggest these as aesthetic embellishments or engagement tactics but rather as epistemic and methodological resources for cultivating different ways of perceiving, relating, and imagining sustainability. We acknowledge that there are multiple potential pathways towards science communication as sustainability, but we believe that arts-based approaches can become particularly salient to our reimagining of science communication as sustainability as they move beyond the 'limited and limiting conceptions of knowledge' prevalent in mainstream science communication (Eisner, 1991/2017, p 2). By arts-based approaches, we refer to methodological tools used by science communication researchers (Leavy, 2020), innovative media and participation formats developed by science communication practitioners, and creative practices through which participants express themselves during science communication initiatives (Nathan et al, 2023). Arts-based approaches, as this volume will demonstrate, offer significant democratizing and inclusive potential, inviting communicators and participants alike towards a fundamentally new mindset.

Arts-based approaches emerged in social sciences and health sciences research during the 1970s, embodying a new ethics of care that redefined the roles of researchers and participants. Within this ethical framework, participants shifted from being mere subjects to becoming collaborators or co-researchers, with enquiry developing through relational and dialogical skills collectively nurtured by both researchers and participants (Finley, 2005). This shift triggered a 'crisis of representation', raising critical questions about the adequacy of traditional quantitative methods for representing and sharing knowledge with an increasingly diverse community of peers (Denzin and Lincoln, 2005). Arts-based methods provided alternative pathways to science's dominant modes of knowing by foregrounding experiential insight, emotional resonance, and empathy – what philosopher K.E. Løgstrup described as a type of recognition distinct from the recognition provided by theories, data, and standards (1976, cited in Damsgaard and Phoenix, 2022). Thus, arts-based research acknowledges multiple ways of knowing (preverbal, sensory, kinaesthetic, imaginary), recognizing art's capacity to reveal truths, foster self-awareness, and generate insights into realities otherwise neglected or dismissed (Gerber et al, 2012; Damsgaard and Phoenix, 2022).

Inspired by this democratizing potential, we aim to cultivate a sustainability mindset that actively embraces imaginative, relational, and participatory engagement. Some of the chapters in the book explore how arts-based forms can inform the design and enactment of science communication practices that foreground the entanglement of human and more-than-human systems and that cultivate more participatory and relational ways of living with sustainability challenges. All chapters, however, challenge the way we have traditionally viewed and done 'effective science communication'. Rather than optimizing messages for uptake, the chapters invite a reorientation of science communication towards world-making practices that attend to relations, responsibilities, and the conditions of collective transformation – an orientation that is central to working with science communication as sustainability.

Reading this book

The chapters in this book can be read separately, as focused investigations, or in sequence. Taken together, they move along two closely connected lines of enquiry: the first three chapters (on the terroir of science communication and the necessity of working with imagination and with emotions) invite science communicators to critically reconsider dominant scientific worldviews and their role in shaping unsustainable relations to the world. This first cluster of chapters directly challenges the universalizing, detached, and instrumental approaches to science communication.

One of the central metaphors explored in this book is 'terroir', as presented in Chapter 2. Here we, together with Thorek Jensen, invite reflection on how science communication emerges from – and is inseparable from – its cultural, social, temporal, spatial, and more-than-human entanglements. With the metaphor 'terroir', we seek to shift science communication from being a universalizing endeavour to an embodied, earthbound one. Brewing on the wine metaphor, the chapter foregrounds the irreducible complexity inherent in communication practices, challenging attempts to reduce effective engagement to predetermined or universally applicable factors. Terroir also asks us to recognize multiple valid ways of knowing, reminding us that communicating science is always shaped by where we are, with whom we interact, and through what stories, bodies, and materials we engage. In doing so, terroir reframes science communication as an emergent, relational practice, cultivated through dynamic interactions between bodies, ecologies, and knowledge. In our practices, it means resisting 'one-size-fits-all' approaches and instead cultivating situated, relational, and more-than-human forms of science communication that foster deeper, sustainable relations.

Imagining sustainable futures requires more than recognizing that our current path is unsustainable – it demands actively seeking perspectives that can expand our imagination and offer fundamentally different visions of living sustainably. In Chapter 3, Achiam turns explicitly to voices and perspectives that have traditionally been marginalized or excluded from dominant narratives about the future, and who, in this way, have already experienced the end of their world. By drawing on feminist and queer theory, anti-colonialist and Indigenous scholarship, and futurist practices, Achiam cultivates diverse and transformative imaginaries that envision entirely different ways of being in and relating to the world, offering her own imaginations of what science communication as sustainability might become.

Expanding our imaginaries is not only a matter of whose perspectives we include but also of how we understand the processes through which imagining and knowing take place. In Chapter 4, Vitting-Seerup demonstrates how emotions are deeply intertwined within both scientific and communicative practices. As sensing, feeling beings, she argues, our ways of knowing extend far beyond intellect alone. Emotions are not separate from our experiences of conducting, engaging with, or interpreting research – they are an integral part of them. This means that recognizing the emotional dimensions of sustainability science, and of science communication more broadly, becomes essential for practising science communication as sustainability. Put simply: emotional investment is not something we add to science communication to better engage an audience. Emotions are already embedded in every question of sustainability science. Acknowledging and foregrounding this is crucial to how we communicate, and imagine, sustainability itself.

In Chapter 5, Grünfeld, together with a cartoonist, a poet, a photographer, and a performance artist, disrupt the book's trajectory by turning to artistic and creative practices. In the chapter, each contributor introduces a piece of work – a poem about microbial entanglements (by Adam Dickinson), a photographic series of lichens (by Alison Pouliot), and a sound recording and performance on fungi (by Maria Brænder) – to enrich the discussion of artistic approaches to science communication. These works open a space for discussing sustainability as a contested and situated concept, the agency of the more-than-human, the transformative force of arts-based approaches, and the performative enactment of knowledge. A fourth work – a comic that thematizes the conversation with artists (by Sofie Louise Dam) – threads through the chapter, weaving together the subtle interplay of humour and seriousness that emerged across the dialogues. The chapter reimagines sustainability communication not as transmission but as participation, attunement, and imaginative responsiveness. Maybe your practice can do the same?

After this artistic disruption, the three final chapters (on making things together, thinking with history, and listening to the messy and unseen) experiment with what becomes possible when science communication is practised through more radically collaborative media and modes. By being in dialogue with artists and other non-scientists, they show how to open space for new relations with the more-than-human world and for the co-creation of ideas, meanings, and knowledge. These chapters move the implications of science communication as sustainability from questions of worldview and imagination into questions of practice: how we work with others, with materials, and with the worlds we seek to engage.

Relating to the world in new ways calls for attending to how we interact with it – quite literally. In Chapter 6, Whiteley explores ideas of relationality by examining how making things together can serve as a generative approach to navigating diverse perspectives and ways of knowing. Drawing on examples from Medical Museion in Copenhagen, Whiteley shows how material co-creation fosters dialogue – not by smoothing over differences but by making space for disagreement, negotiation, and shared meaning making. In doing so, Whiteley positions collaborative making as both a method and a metaphor for science communication as sustainability: rooted, embodied, open-ended, and kind.

Where Whiteley foregrounds relationality through shared, material practices in the present, Thorek Jensen extends this relational work across time. In Chapter 7, Thorek Jensen introduces thinking with history as a future-thinking practice: an approach that looks beyond the constraints of the present by engaging the past as a generative resource for imagining alternative sustainable futures. Approaching sustainability in this way requires drawing on diverse ideas, perspectives, experiences, and conceptual frameworks in order to expand what futures can be imagined and shaped. Crucially, the invitation is not to repeat the past but to think with it. Thorek Jensen thus frames sustainability as a call for epistemic and cultural renewal – one that opens new trajectories and more inclusive imaginaries of what the future might become. For science communication practices, this entails becoming attentive to the historical ideas, assumptions, and power relations we think with and critically mobilizing the past not merely to account for where we come from but as thinking material for shaping imaginaries of more sustainable futures.

If thinking with history invites us to rework the temporal horizons of science communication, the final chapter extends this relational reorientation into the sensory domain. Chapter 8 moves our exploration into a deeper sensory engagement by focusing on sound as a transformative arts-based method in science communication. Drawing on historical whale research and Danish sound artist Maria Brænder's work with zooplankton, Grünfeld demonstrates how listening deeply connects us to hidden and overlooked worlds of other organisms. More than just focusing on sound, however, the chapter emphasizes listening as an embodied, relational practice that can expand our senses and challenge dominant visual and discursive frameworks within science communication. You are even invited into this exploration experientially, by engaging with audio materials that amplify imagination and foster transformative experiences throughout the chapter and its many links to audio files.

Together, these chapters do not offer a single method for working with science communication as sustainability but rather propose a repertoire of sensibilities for sensing, imagining, and enacting more liveable relations with human and more-than-human worlds. We hope you take what you need and create what you lack from here; the work has just begun.

References

Arrhenius, S., and Holden, E.S. (1897) 'On the influence of carbonic acid in the air upon the temperature of the Earth', *Publications of the Astronomical Society of the Pacific*, 9(54): 14–24.

Austin, J.L. (1963) 'Performative-constative', in C.E. Caton (ed) *Philosophy and Ordinary Language*, University of Illinois Press, pp 22–54.

Bal, M. (2009) *Travelling Concepts in the Humanities: A Rough Guide*, University of Toronto Press.

Beck, U. (2016) *The Metamorphosis of the World*, Polity.
Bonneuil, C., and Fressoz, J.-B. (2017) *The Shock of the Anthropocene: The Earth, History and Us* (D. Fernbach, trans), Verso. [Original work *L'événement Anthropocène: La terre, l'histoire et nous* published in 2013.]
Brundtland, G.H. (1987) 'Report of the World Commission on Environment and Development: Our common future', Available from: https://www.brundtland.co.za/wp-content/uploads/2022/08/Brundtland-Report-1987-Our-Common-Future.pdf
Bucchi, M., and Trench, B. (2021) 'Introduction: Science communication as the social conversation around science', in M. Bucchi and B. Trench (eds) *Routledge Handbook of Public Communication of Science and Technology* (3rd edn), Routledge, pp 1–13.
Canon, C.R., Boyle, D.P., and McAfee, S.A. (2022) 'Visualizing the structure and development of climate change communication research', *Journal of Science Communication*, 21(7): A03.
Clark, W.C. (2007) 'Sustainability science: A room of its own', *Proceedings of the National Academy of Sciences*, 104(6): 1737–8.
Damsgaard, J.B., and Phoenix, A. (2022) 'World of change: Reflections within an educational and health care perspective in a time with COVID-19', *International Journal of Social Psychiatry*, 68(1): 177–82.
Denzin, N.K., and Lincoln, Y.S. (2005) 'Introduction: The discipline and practice of qualitative research', in N. K. Denzin and Y.S. Lincoln (eds) *Sage Handbook of Qualitative Research* (3rd edn), Sage, pp 1–32.
Dürr, H.-P., Dahm, D., and zur Lippe, R. (2005) 'Potsdam Manifesto 2005: "We have to learn to think in a new way"', Federal Ministry of Education and Research of Germany.
Eisner, E.W. (1991/2017) *The Enlightened Eye: Qualitative Inquiry and the Enhancement of Educational Practice*, Macmillan.
European Commission (2022) 'GreenComp, the European sustainability competence framework', Joint Research Centre, Publications Office of the European Union, Available from: https://data.europa.eu/doi/10.2760/13286
Finley, S. (2005) 'Arts-based inquiry: Performing revolutionary pedagogy', in N.K. Denzin and Y.S. Lincoln (eds) *Sage Handbook of Qualitative Research* (3rd edn), Sage, pp 681–94.
Fischer, D., Lüdecke, G., Godemann, J., Michelsen, G., Newig, J., Rieckmann, M. et al (2016) 'Sustainability communication', in H. Heinrichs, P. Martens, G. Michelsen, and A. Wiek (eds) *Sustainability Science: An Introduction*, Springer Netherlands, pp 139–48.
Fourier, J.-B.J. (1837) 'General remarks on the temperature of the terrestrial globe and the planetary spaces', *American Journal of Science*, 32(1): 1–20.
Funtowicz, S.O., and Ravetz, J.R. (1993) 'Science for the post-normal age', *Futures*, 25(7): 739–55.

Gerber, N., Templeton, E., Chilton, G., Liebman, M.C., Manders, E., and Shim, M. (2012) 'Art-based research as a pedagogical approach to studying intersubjectivity in the creative arts therapies', *Journal of Applied Arts & Health*, 3(1): 39–48.

Godemann, J., and Michelsen, G. (2011) 'Sustainability communication: An introduction', in J. Godemann and G. Michelsen (eds) *Sustainability Communication: Interdisciplinary Perspectives and Theoretical Foundation*, Springer, pp 3–11.

Heidegger, M. (1968) *An Introduction to Metaphysics*, Yale University Press.

Hochschild, A.R. (1979) 'Emotion work, feeling rules, and social structure', *American Journal of Sociology*, 85(3): 551–75.

Horst, M., and Michael, M. (2011) 'On the shoulders of idiots: Re-thinking science communication as "event"', *Science as Culture*, 20(3): 283–306.

Leavy, P. (2020) *Method Meets Art: Arts-Based Research Practice* (3rd edn), Guilford Press.

Løgstrup, K.E. (1976) *Kunst og erkendelse* [Art and cognition], Udgivelse.

Loura, I.C.D., and Dickinson, R. (2016) *Fractal Sustainability: A Systems Approach to Organizational Change*, Routledge.

Lyle, J. T. (1996) *Regenerative Design for Sustainable Development*, John Wiley & Sons.

Margolis, J. (1957) 'Notes on the logic of simile, metaphor and analogy', *American Speech*, 32(3): 186–9.

McLuhan, M., and Fiore, Q. (1967) *The Medium is the Massage: An Inventory of Effects*, Random House.

McNeill, J.R., and Engelke, P. (2014) *The Great Acceleration: An Environmental History of the Anthropocene since 1945*, Harvard University Press.

Mochizuki, Y., and Yarime, M. (2016) 'Education for sustainable development and sustainability science: Re-purposing higher education and research', in M. Barth, G. Michelsen, M. Rieckmann, and I. Thomas (eds) *Routledge Handbook of Higher Education for Sustainable Development*, Routledge, pp 11–24.

Nathan, S., Hodgins, M., Wirth, J., Ramirez, J., Walker, N., and Cullen, P. (2023) 'The use of arts-based methodologies and methods with young people with complex psychosocial needs: A systematic narrative review', *Health Expectations*, 26(2): 795–805.

Nielsen, K. (2013) 'Scientific communication and the nature of science', *Science & Education*, 22(9): 2067–86.

Roös, P.B. (2021) *Regenerative-Adaptive Design for Sustainable Development*, Springer International Publishing.

Stengers, I., and Muecke, S. (2018) *Another Science is Possible: A Manifesto for Slow Science*, Polity.

Sterling, S. (2004) 'Higher education, sustainability, and the role of systemic learning', in P.B. Corcoran and A.E. Wals (eds) *Higher Education and the Challenge of Sustainability*, Springer, pp 49–70.

Takach, G. (2016) *Scripting the Environment: Oil, Democracy and the Sands of Time and Space*, Palgrave Macmillan.

Van der Ryn, S., and Cowan, S. (2013) *Ecological Design*, Island Press.

2

We Need to Terroirize Science Communication for It to Be Sustainable: A Situated, Embodied, and Relational Practice

Martin Grünfeld, Marianne Achiam, Jacob Thorek Jensen, and Sabrina Vitting-Seerup

What if we did not treat science communication as a neutral and objective practice? What if we treated it as something deeply shaped by place, history, environment, culture, and actors (both human and more-than)? What if we develop our approach to science communication *as* sustainability through the metaphor of *terroir*?

In this chapter, we explore how terroir, used as a metaphor and analytical orientation, can help articulate a pluralistic and situated approach to science communication that foregrounds the material, ecological, social, and historical conditions under which knowledge is communicated and made meaningful (see also Achiam et al, 2025). We write from within contemporary Western science communication research, addressing readers situated in similar institutionalized traditions, many of whom are already aware of the limitations of linear, deficit-oriented, and universalizing models. Our aim is therefore not to rediscover relational or situated knowledge practices, which have long histories across diverse cultures and knowledge traditions (Orthia, 2020; Duan et al, 2023; Kankaria et al, 2023) but to propose terroir as a conceptual device that helps reconfigure how such concerns are recognized, analysed, and acted upon within dominant science communication frameworks.

Through the metaphor of terroir, we propose an approach to science communication as sustainability that is grounded in situated interventions and in the enactment of coupled networks of ecological and social, human and non-human, micro and macro actors. By developing terroir from metaphor towards theory, we explore how terroirizing science communication can shift both analytical attention and communicative practice, especially in contexts where sustainability challenges expose the limits of universal solutions and decontextualized expertise.

We have deliberately verbalized the noun terroir in this chapter's title. This provocation builds on our earlier work, where we introduced terroir as a conceptual metaphor for understanding science communication as a situated, relational, and material practice (Achiam et al, 2025). While terroir as a noun can function as a descriptive-analytical framework for reassessing science communication events, *terroirizing* science communication is intended to do more than describe. It sets out a normative orientation and a call to action: to attend explicitly to the ecological, social, historical, and ethical conditions that shape science communication and to take responsibility for how those conditions are enacted and reproduced. In this sense, terroirizing operates as a form of reparative critique that calls us to do more in a time of crisis (Sedgwick, 1997; Latour, 2004). In this chapter, we extend this conceptualization by explicitly situating our terroirizing of science communication within debates on sustainability, more-than-human relations, and arts-based science communication.

Terroir is commonly used to describe the environmental and cultural conditions that contribute to the distinctive qualities of wines from particular regions – conditions that emerge from complex relations among geology, climate, soil, microbes, plants, animals, human labour, traditions, technologies, and temporal rhythms (van Leeuwen, 2009; Jayer, 2018; Kramer, 2018). Importantly, terroir does not reside in any single factor but in the dynamic relations among them. In this way, it resonates with long-standing critiques – both within and beyond Western scholarship – of the separation between nature and culture, knowledge and place, experts and environments. When we draw on terroir here, we do so in dialogue with these broader traditions, not as their replacement.

In the following sections, we extend this relational logic metaphorically from vineyards and soil microbes to science communication practices concerned with sustainability. We begin by framing terroir as a relational and sustainability-oriented mode of knowing, before turning to questions of publics and more-than-human inclusion, place, space, and temporality. We conclude by considering how arts-based approaches may help cultivate a situated and earthbound science communication practice. Terroirizing science communication may feel dizzying – or even uncomfortable – but we argue that such discomfort is productive if science communication is to respond meaningfully to the ecological and social crises of our time.

Terroirizing sustainability science, or how the relational knowledge of the vineyard unfolds

Terroir is a relational way of knowing. As a concept, it 'offers a theory of how people and place, cultural tradition and landscape ecology, are mutually constituted over time' (Paxson, 2010, p 444). It invites a bio-regionalist approach to knowledge that foregrounds relations among territory, craft, knowledge, multisensory aesthetics, and multispecies relationships (Rigobello and Evans, 2024). Such relational epistemologies are not new, nor unique to Western agrarian contexts; they resonate with long-standing knowledge traditions that refuse sharp separations between nature and culture, practice and value (see also Chapter 3). What terroir offers in this context is a particularly well-articulated example of how such relational knowing operates within domains often dominated by modern scientific rationalities.

Terroir is a complex (if not impossible) object to measure, predict, and control. Although it depends on measurable macro-environmental factors such as precipitation, temperature, elevation, and sunlight, it cannot be reduced to them. This has led some scientists to question whether terroir exists at all, or whether it is merely a social construction (Teil, 2012). Yet, for producers, distributors, and consumers, terroir is experienced as real and fundamentally material. As the winegrower Didier Dagueneau expresses it:

> I am deeply attached to the earth. There must be osmosis between man and his terroir; to express this well, he must know how to tune into his soil, be sensitive to it, train all the fibers of his being to sniff the air, to see if the vine is suffering, if the soil is breathing. (Dagueneau, 2018, p 17)

By definition, osmosis refers to the movement of substances across a semipermeable membrane (Gregersen, nd). In the context of terroir, osmosis describes an intimate epistemic relationship constituted through flows, exchanges, and shared vulnerability – what Haraway (2008) would call kinship and shared suffering. Terroir knowledge is embedded in material practices and ways of life, inseparable from ethical and aesthetic commitments. Cultivating terroir involves a non-interventionist approach that allows the land to express itself through taste and sensorial experience, often rejecting synthetic inputs that would abstract flavour from place (Teil, 2012). In that sense, terroir constitutes a relational and sustainability-oriented way of knowing rather than merely a set of experiences.

Such relational knowledge does not sit easily within positivist scientific regimes that assume objects of knowledge to be independent, stable, and fully representable as data (Teil, 2012). Likewise, the sustainability challenges we face today resist mechanistic approaches that fragment the world into discrete, controllable components (Dürr et al, 2005; Benessia et al, 2012). Sustainability science has emerged as a heterogeneous field that acknowledges the entanglement of knowledge, ethics, and lived experience (Kauffman, 2009; Lang et al, 2012), operating under conditions of unpredictability, incomplete control, and a plurality of legitimate perspectives (Funtowicz and Ravetz, 1993). Much like a winemaker, sustainability scientists must cultivate sensitivity to complexity and uncertainty, remaining responsive to the particularities of place, people, and more-than-human relations.

To engage meaningfully with places, communities, and ecologies, sustainability science cannot treat science communication as an add-on (Achiam and Irwin, 2026). Yet within dominant institutional frameworks, science communication is often positioned as something that happens *after* knowledge production, oriented towards informing or persuading external audiences. However, as Lewenstein (1995, 2007) and others have shown, scientific practice is already embedded within networks of public communication (Halpern and Rogers, 2021). From this perspective, science communication is part of research-in-the-making, not merely its dissemination.

For those of us working within Western, institutionalized traditions of science communication, terroirizing science communication therefore calls for a reflexive stance that acknowledges our own situatedness within the socio-ecological problems we study and communicate. Rather than positioning ourselves as external observers or neutral intermediaries, we must recognize how we are already implicated: co-fermenters rather than detached analysts. As Dagueneau remarks: 'Before becoming a winegrower, I saw myself as a farmer' (Dagueneau, 2018, p 17). In a similar spirit, terroirizing science communication invites those of us trained as science communicators to reconsider professional identities, not as neutral conveyors of scientific results but as situated practitioners entangled in the ethical, ecological, and political conditions of sustainability.

Terroirizing publics, or why winegrowers cannot make wine alone

No vineyard exists without human intervention. Vines are naturecultures: shaped through selective cultivation, intergenerational practice, and socio-technical arrangements that have produced varieties that do not exist in the wild (van Leeuwen, 2009). Terroir itself is historically embedded, inseparable from the socio-economics, politics, and techniques of growing and making wine (Seguin, 1986). Importantly, terroir is not produced by a single expert actor but by a distributed collective of growers, pickers, makers, traders, and consumers whose situated knowledges co-constitute what wine becomes.

A similar distinction is necessary when considering science communication. While institutionalized science communication – as a professional field shaped by modern science, media systems, and policy contexts – has a relatively recent history (Davies and Horst, 2016), practices of communicating, negotiating, and contesting knowledge about the natural world are ancient, plural, and globally diverse (Orthia, 2020; Duan et al, 2023; Kankaria et al, 2023). Recognizing this distinction matters because sustainability science communication cannot be understood or transformed without acknowledging the long-standing knowledge practices that predate and exceed Western scientific institutions.

Wicked sustainability problems affect all of us, but not in the same ways. Addressing them means leveraging the collective capacities of society (Canfield et al, 2020), yet science communication, particularly in its dominant Western institutional forms, has historically failed to prioritize epistemic diversity (Dawson, 2014). The challenge is therefore not merely one of reaching diverse audiences but of recognizing diverse publics as knowledge holders and knowledge producers, whose understandings of environmental change are grounded in lived practice and often intergenerational relations with specific places.

Sustainability issues cannot be addressed as discrete objects studied from a distance. They demand engagement with the situated knowledges of those already living with environmental disruption – knowledges that are sensory, affective, embodied, and deeply practical (Achiam et al, 2021; Verlie, 2022). This position builds on long-standing critiques within science and technology studies, which have shown how the separation of scientific expertise from public knowledge obscures the situated, normative, and political dimensions of knowledge making (Wynne, 1992; Jasanoff, 2005). In this sense, an extended peer community is not an add-on to sustainability science communication but its necessary condition (Block et al, 2018).

Our own research illustrates how such reconfigurations can unfold in practice. In 2023, Achiam, Vitting-Seerup, and Whiteley hosted three co-creation workshops in Copenhagen, Denmark (Achiam et al, 2024). Designed as meta-communicative spaces, the workshops brought together researchers, activists, artists, students, consultants, and journalists to collectively reflect on sustainability science communication. When participants were invited to adapt a visual map of the field, the exercise quickly exceeded its initial framing: some folded or reshaped the maps, others requested materials to add depth, while discussions questioned whether cartography, or even ecosystem metaphors, were adequate representations at all. These interventions did not simply enrich the communicative process; they actively reshaped what counted as sustainability science communication, including which actors, relations, and forms of agency were rendered visible. One participant asked whether a tree might communicate sustainability science, prompting another to reply, laughing, 'Perhaps better than a member of parliament'.

Terroirizing publics, therefore, means attending not only to who is included but to how roles, authority, and agency are distributed within communicative processes. This includes making space for voices across political positions, income levels, educational backgrounds, and minoritized and marginalized communities while remaining open to the unexpected trajectories that collective communication events may take (Horst and Michael, 2011). Importantly, terroirizing science communication does not claim originality in its call for inclusion. Rather, it seeks to synthesize long-standing critiques of epistemic hierarchy, extractive engagement, and Western dominance by offering a metaphor that foregrounds relationality, situatedness, and mutual dependence.

Borrowing terroir from the wine industry must not mean importing its inequities. Wine production has long relied on exploited labour (Cerillo, 2024), and terroirizing science communication explicitly rejects such extractive arrangements to include marginalized voices. Science communication as sustainability is conceived here as a radically collectivist endeavour, in which diverse publics, researchers, and stakeholders share epistemic responsibility and ownership (for more on this, see also Rasekoala, 2023). Knowledge is not disseminated outward from a privileged centre but emerges through inclusive, place-attuned, and ethically grounded practices oriented towards remaking futures together.

Terroirizing inclusion, or why humans cannot make wine by themselves

What do a tree, a door, or a microbe do in a science communication event? And what do we do with (or to) them as we plan, execute, and conclude such events? In the spirit of disenchanted modernity, it may be tempting to dismiss these questions as childish play or to meet them with laughter. Yet as Sigmund Freud suggested, our enjoyment of a joke often signals what is repressed in more serious conversation (Billig, 1997). Taken seriously, such questions draw attention to the role of non-human actors in science communication.

Engaging with neglected non-human members of the terroir requires us to unlearn deeply ingrained forms of anthropocentrism that continue to shape Western traditions of science communication. Rather than treating non-humans as passive backdrops or symbolic props, terroirizing includes asking how trees, doors, microbes, and other material actors participate in shaping what can be sensed, known, and communicated. As Natasha Myers suggests in her exploration of old-growth Black Oak trees, this may require 'forget[ting] everything we thought we knew about non-human lives and worlds' in order to cultivate an ethic of humility and attentiveness towards the unknowable and imperceptible (Myers, 2020, p 75). Such an ethic does not romanticize non-humans but recognizes them as active participants in landscapes that register and respond to environmental change over time.

From this perspective, terroirizing science communication involves troubling the assumption – still implicit in many definitions – that science communication is primarily a social exchange among humans (Bucchi and Trench, 2021). Terroir instead foregrounds the material-semiotic relations through which communication events are constituted, including relations with non-humans that shape what becomes possible to perceive, articulate, and contest. This is not a call to 'add' trees or microbes as decorative elements but to acknowledge how communicative practices are already co-produced through human–non-human entanglements.

The wine metaphor makes this clear. Beyond the intimate relationship between a winemaker and a vineyard lies an intricate microbial terroir: dynamic co-associations among bacteria, fungi, plants, and soils that shape vine health and wine quality (Gilbert et al, 2014). The common grapevine (*Vitis vinifera*) is inseparable from these microbial relations. Attending to such entanglements shifts attention away from individual agency towards distributed processes through which material conditions actively configure outcomes. Similarly, in science communication, non-humans do not merely enrich experience; they constrain, afford, and redirect communicative possibilities, shaping what knowledge can emerge and how it circulates.

While a terrorization of science communication involves a self-reflexive engaged communicator and includes a diversity of people, we also need to cultivate a more expansive non-anthropocentric understanding of inclusion. Perhaps an *art of inclusion*, as suggested by Anna Tsing (2010), that promotes a passionate immersion into the lives of other species. Such an art does not seek to map or instrumentalize the terroir, nor to replace scientific analysis with critique, but to hold cultural and scientific practices together in ways that remain responsive to more-than-human relations. This stance invites curiosity towards what non-humans can do in science communication events, not in the service of productivity but in their capacity to act differently and to trouble dominant modes of knowing.

Instead of treating non-human actors as measurable variables, we must approach them as variations through an art for noticing that takes seriously the roles doors, trees, and microbes may play in the terroir-experience of a science communication event. Importantly, this does not imply an ontological levelling (or flattening) of all actors participating in the terroir. Terroirizing inclusion does not claim that doors, microbes, and humans act in the same ways and carry equal responsibility. Rather, it foregrounds how bodies, built environments, and ecosystems are differentially entangled through economic, political, and environmental relations (Alaimo, 2014).

To terroirize science communication, then, is to move beyond human exceptionalism and to attend to how human and non-human actors jointly shape the situated conditions under which sustainability knowledge is communicated, contested, and reimagined.

Terroirizing place and space, or how the environment is (at least part of) the message

Why mention a door alongside a microbe and a tree? It might seem more intuitive to distinguish between living organisms and inert material objects when considering the terroir as a dynamic set of relations. Yet the vineyard itself is an entangled natureculture that includes not only fields, plants, and farmers but also built structures such as farmhouses, cellars, and wooden barrels. Barrels, for example, participate in the winemaking process by hosting microbial relations that shape ageing and flavour profiles, relations that are further influenced by wood type, cellar environments, and practices of cleaning and gatekeeping (Rigobello and Evans, 2024). Even a door plays a role in these dynamics by regulating exchanges between inside and outside, enabling wine to mature under specific conditions. As Bruno Latour, writing under the pseudonym Jim Johnson, famously noted, without doors there would simply be holes in walls allowing anything to enter or escape (Johnson, 1988). His humorous example underscores a serious point: material objects participate in shaping social practices.

Attending to the terroir of science communication likewise requires attention to how environments, physical structures, and ecologies co-produce communication events. As geographers remind us, 'everything happens somewhere' (Agnew, 2011). Communication does not take place in neutral containers; locations actively shape how the communication unfolds (Crang and Thrift, 2000; Shome, 2003). These locations can be understood in terms of both place and space. Places such as museums, vineyards, or laboratories are sites imbued with meaning through human practices and histories (Cresswell, 2015). Spaces, by contrast, refer to dynamic and relational configurations – such as weather systems, atmospheres, or climatic conditions – that are continually in flux (Massey, 2012). While the distinction between place and space has long been debated (Agnew, 2011; Agnew and Coleman, 2018), holding both concepts in view allows us to attend simultaneously to meaning, movement, stability, and change in science communication.

Recognizing the terroir of a science communication event thus involves attending to how physical, material, environmental, atmospheric, and acoustic conditions shape what can be sensed, expressed, and contested. Importantly, places and spaces are never fixed. Vineyards, like museums or outdoor communication sites, are shaped through seasonal variation, movement, and embodied inhabitation. Environments are continuously formed through movement and embodiment intersected or intermeshed by multiple bodies visiting, resting, and leaving (Ingold, 2021). In this sense, places are stabilized precisely through ongoing flux and exchange.

While human practices such as curation, design, and facilitation clearly shape science communication settings, they do so in concert with non-human forces. In vineyards, microbes, weather patterns, and climatic conditions often exert greater influence over spatial transformation than human intention alone. Similarly, in science communication, non-human elements such as materials, infrastructures, and atmospheres actively condition how events unfold and how they are experienced. Spaces and places also act upon bodies. Architectural forms, landscapes, and atmospheres affect how communication is felt, not only how it is understood. As Peter Sloterdijk has shown, built environments can shape collective affect and bodily orientation through acoustics and spatial design (Sloterdijk, 2008). Just as geology, topography, and microbial life shape the terroir of wine, material environments shape how sustainability issues are encountered, interpreted, and inhabited in science communication.

This, however, does not imply environmental determinism. Rather, the environment is a core part of the message, just as terroir is an expression of place, yet the message remains co-constituted by communicative practices, embodied and affective experiences, and the material-spatial conditions in which they are embedded.

Terroirizing science communication, therefore, involves recognizing how places and spaces participate in shaping not only how communication happens but what can be communicated at all. As communication unfolds in a mesh of interconnections (Morton, 2010), neither message nor communicator occupies a privileged position. A place may reinforce a communicative aim or undermine it. Moreover, just as winemaking reshapes the land it depends on, science communication practices also leave material and ecological traces. This is particularly salient in sustainability contexts: the setting is (at least part of) the message. Science communication as sustainability is thus not only a radically collective more-than-human endeavour but an earthbound practice whose material conditions and consequences cannot be abstracted away.

Terroirizing temporalities, or how plural rhythms and situated time shape science communication

'Time is our most important winemaking tool' (Boiling, 2020). The German Riesling producer Gut Hermannsberg recently postponed the release of its top Rieslings to five years after harvest in order to allow the terroir to fully unfold. Patient waiting is, however, only one among multiple temporal dimensions shaping wine. Terroir emerges through the interplay of diverse rhythms: biological growth cycles, seasonal climatic patterns, microbial dynamics, and slow geological processes, interwoven with periodic human interventions ranging from harvesting and fermentation to long-standing cultural practices and collective memory (van Leeuwen, 2009; Johnsson, 2015; Phillips, 2020). Terroir is thus not only spatially situated but temporally constituted through relations unfolding across multiple timescales.

Science communication, by contrast, is often organized around implicit assumptions of linear, universal time. To terroirize temporality is to resist such flattening by grounding time in the lived, situated, and more-than-human contexts through which knowledge unfolds. Rather than treating time as a neutral backdrop, terroirized science communication attends to how biological, geological, cultural, and institutional temporalities become entangled in specific communicative practices (see more in Chapters 3 and 7). In this sense, science communication mirrors sustainability itself, which is governed by an 'irreducible temporal complexity' – a plurality of rhythms and periodizations that shape how progress, urgency, and uncertainty are understood (Vostal et al, 2019, p 799). Terroirized science communication makes visible how sustainability is negotiated through time, between what can be known now, what has been forgotten, and what must be imagined.

Sustainability science further demands attentiveness to temporal complexities. Ecological cycles, cultural memory, and institutional rhythms often operate out of sync (Felt, 2024). While scientific understanding of urgent problems depends on both retrospective awareness and anticipatory futures (Evans and Achiam, 2021), disparities in temporal scale – such as between rapid soil degradation and slow regeneration (Held, 2001) or fossil fuel consumption and their geological formation (McNeill and Engelke, 2014) – complicate the alignment of knowledge, policy, and practice. These mismatches do not merely delay action; they shape which forms of knowledge gain traction and which are marginalized.

Building on Felt's (2024) work, terroirizing temporality also foregrounds the uneven temporal capacities of different actors to engage in science communication. As Felt notes, participation is not only a matter of access or motivation but of temporal alignment: the ability to inhabit, navigate, and negotiate the temporal structures embedded in scientific, political, and everyday life. Scientists, policy makers, communicators, and publics operate within different rhythms, time horizons, and degrees of temporal control, shaping how, when, and even *whether* engagement becomes possible. A terroir approach emphasizes these asymmetries by acknowledging that the 'right moment' for engagement is rarely shared, and that meaningful engagement must therefore be processual and temporally plural.

Crucially, terroirized science communication also engages with more-than-human temporalities – what Howland (2025) refers to as nature time. Migratory cycles, erosion, growth, decay, and regeneration are not passive backdrops to human action but active forces shaping what sustainability means and how it unfolds. Attuning to these temporalities requires moving beyond anthropocentric frameworks towards practices responsive to the temporal signatures of specific ecosystems. In doing so, science communication as sustainability can bridge disciplinary and social divides, grounding action in the material and temporal realities of place.

Terroirizing temporality thus reimagines science communication as a sustained practice attentive to temporal diversity. It prompts questions such as: who has time to participate? Whose rhythms are prioritized in engagement design? How might communicative environments accommodate slow contemplation, cyclical return, and episodic urgency alike? Rather than treating publics as passive recipients of accelerated technoscientific futures or calling for a blanket 'slowing down' of science, terroirized science communication cultivates temporal solidarities across differences, attuning to both human lives and more-than-human rhythms. In this sense, temporality becomes an integrative dimension of terroir: a way of situating knowledge not only in space and relation but in time, enabling science communication practices that enact sustainability rather than merely describe it.

Earthbound, or how to cultivate the terroir of science communication through arts-based approaches

As we unravel terroir as a metaphor for science communication as sustainability, its multifaceted character becomes increasingly apparent. While terroir refers to a complex set of relations, it is often experienced in seemingly simple encounters between wine, glass, and a tasting subject. For most people, terroir manifests as a distinctive gustative quality (Teil, 2012). However, even when wine is poured into a glass and tasted, its 'message' cannot be contained. Wine cannot hold its 'message'; it leaks outward, implicating temporalities of soil formation, microbial life, craftsmanship, storage, and circulation. Like a leaky amphora, terroir disrupts linear transmission models of communication and pushes science communication beyond container metaphors, recognizing that all participants do not experience or interpret these relations in the same way.

Yet to experience the terroir of a wine requires an acquired taste. This simple point emphasizes a potential limitation for our reframing of science communication as sustainability through the metaphor of terroir: if experiencing terroir depends on expertise, cultivated sensory capabilities, and relational knowledge, does terroirized science communication risk becoming exclusive?

To respond to this concern and unsettle the potentially euphemized connotations of terroir, we turn to Jeff VanderMeer's science-fiction masterpiece the *Southern Reach Trilogy*. Set in the unsettling wilderness of Area X, the trilogy explores an irreducible entanglement of place, consciousness, and ecosystem. To express this entanglement, the notion of terroir emerges in the trilogy to describe a force that radically transforms human characters, dissolving the distance between observer and environment (see also Sell, 2018). Area X is not inclusive by invitation but by implication: every single organism is already embedded and affected. This makes the trilogy analytically useful, not as a sustainability blueprint but as a speculative device for thinking through what it means to lose epistemic distance in the Anthropocene. Here, there is no outside position from which to observe the sustainability crises without being implicated in them. Science communication as sustainability, therefore, always already involves affected publics.

In the Anthropocene, we live in the middle of disaster (although it is experienced in various degrees) – a place not that different from Area X calling for us to respond and adapt. A core question for us to consider, then, is how to cultivate adaptive ways forward in a time of crisis? As Latour and Peter Weibel argue, contemporary societies lack not only technical solutions but imaginative and emotional capacities to grapple with existential planetary challenges, calling for aesthetic practices that render us sensitive to other forms of life and emerging futures (Latour and Weibel, 2020). Where scientific language often abstracts and distances, arts-based approaches can cultivate collective sensibilities for knowing and imagining otherwise (Leavy, 2020). They do so without replacing scientific knowledge or political action but by complementing and reorienting them. Arts-based approaches support this reorientation by engaging affect, perception, and embodiment. By evoking emotion (Eisner, 2008), opening new ways of seeing (Cooke et al, 2017), and inviting embodied reflection (Taylor and Ladkin, 2009), such approaches can help cultivate the longing for change that often precedes transformation. This longing is not a guarantee of action but a condition for rethinking how knowledge and futures are related. In this sense, arts-based science communication does not aim to simplify sustainability but to make its entanglements experientially graspable.

Terroirizing science communication thus calls for an earthbound practice attentive to medium, format, place and space, participants (human and non-human), and temporal plurality. It rejects the illusion of a neutral vantage point from which sustainability issues can be communicated without consequence. Instead, it acknowledges the messiness of entangled crises – climate change, war, inequality – and calls for new forms of proximity to the situations we inhabit. Cultivating such proximity requires arts-based approaches but also the unlearning of aesthetic distance, particularly in institutional settings such as museums, where detachment is often normalized (Ziebritzki, 2020). If the relatively brief history of science communication has focused on optimizing transmission and measuring effectiveness, its future may need to rebalance this trajectory by developing an art of science communication that complements analytical rigour with situated, sensory, and ethical engagement.

Concluding remarks: The many truths of terroir

If our proposal to terroirize science communication sounds messy, extensive, and uncontrollable, it is because it is meant to be. In an era marked by ecological crisis and epistemic uncertainty, it would be far more troubling to continue to rely on linear, reductive modes of thinking that have long contributed to environmental degradation by separating nature from culture and knowledge from consequence. Terroir, as we have developed it here, is not a call for conceptual neatness but for reimagining science communication in its irreducible multiplicity.

Introducing terroir allows us to understand communication as always already entangled: social and ecological, situated yet globally connected, dynamic and relational, multi-factorial yet resistant to reduction into single measurable variables. Terroir offers a way of holding together science, practice, and art without collapsing their differences. As a metaphor, it resists both linear transmission models and simplified dialogical ideals, not by replacing them with another model but by opening science communication to forms of engagement that foreground transformation rather than transmission.

This reorientation also requires us to reconsider our roles as researchers and science communicators. Terroir draws attention to how places, spaces, bodies, and temporalities co-produce communication events in ways that defy neat separations between sustainability, science, politics, ethics, and aesthetics. To terroirize science communication is therefore not to master complexity but to tend it by remaining attentive to the many forms of knowledge that emerge through situated relations.

Terroirized science communication requires living with uncertainty without sliding into apathy or relativism. Uncertainty does not imply that nothing matters; on the contrary, it reminds us that many things matter, often simultaneously. It invites experimentation, attentiveness, and a willingness to adjust – to tweak, to play – even when outcomes cannot be fully predicted or explained. As Dagueneau (2018) notes, there is not one truth of the vineyard, but several. This multiplicity does not signal the end of science but the possibility of forms of enquiry that bring us closer to complex truths through multiple ways of knowing.

Throughout this chapter, we have explored how humans and non-humans as well as places and spaces and plural temporalities shape the terroir of science communication. We have argued that cultivating this terroir calls for arts-based approaches capable of working across science, politics, ethics, and aesthetics. These approaches do not promise solutions, but they can create conditions for ethico-aesthetic (Guattari, 1995) encounters that bring bodies, affects, and lived experience into communication events alongside the multiple agencies involved in sustainability challenges. Yet terroir is always more than we anticipate. What we have outlined here is therefore not a programme but a point of departure.

Implications for science communication as sustainability

The metaphor of terroir offers a particularly productive frame for seeing how sustainability involves situated relational knowledges with ethical and aesthetic dimensions. As Adrien Rigobello and Joshua Evans (2024) note, experiencing the terroir of wine can draw drinkers into ethical relations with the more-than-human processes involved in its making. Similarly, aesthetic encounters with the terroir of science communication may help cultivate an ethics of deep sustainability grounded in proximity rather than distance. Yet the Anthropocene is disorienting and existentially demanding. As Latour and Weibel suggest, in such conditions we must find ways to 'land' without assuming a stable ground (Latour and Weibel, 2020). The earth we inhabit – and that inhabits us – is complex, uneven, and often uncomfortable. Terroir invites us to stay with this discomfort, to relinquish the fantasy of distance, and to reimagine science communication as sustainability as an earthbound practice: situated, responsive, and resonant with the many truths through which futures are made.

When working with the metaphor of terroir, many teachings of winemaking lead to new questions for science communication research and practice:

- *Winegrowers cannot make wine alone*
  How can we include more people with different backgrounds and perspectives in the co-creation of our science communication events?
- *Humans cannot make wine by themselves*
  How can we start to honour the capabilities that non-humans possess in our science communication events instead of trying to ignore or control them?
- *The environment is part of the wine*
  How do we make space for the environment to express itself in sustainable ways instead of only discussing it in distancing language?
- *Winemaking is dynamic, rhythmic, cyclical*
  How do we include other rhythms and cycles in our science communication events?
- *Cultivating a terroir is an earthbound practice*
  How do we develop science communication that attends to the many truths of sustainability science without resorting to nihilism or relativism?

References

Achiam, M., and Irwin, A. (2026) 'Post-normal science communication? Evidence of third-order thinking among sustainability scientists', *Public Understanding of Science*, 35(2): 159–76.

Achiam, M., Dillon, J., and Glackin, M. (eds) (2021) *Addressing Wicked Problems through Science Education: The Role of Out-of-School Experiences*, Springer.

Achiam, M., Vitting-Seerup, S., Whiteley, L., and Dam, S.L. (2024) 'Creating resonance with arts-based approaches to sustainability science communication', *Journal of Science Communication*, 23(6).

Achiam, M., Grünfeld, M., Vitting-Seerup, S., Jensen, J.T., and Whiteley, L. (2025) 'Towards a terroir approach to science communication and its evidencing', *Journal of Science Communication*, 24(7).

Agnew, J. (2011) 'Space and place', in J. Agnew and L. David (eds) *Handbook of Geographical Knowledge*, Sage, pp 316–30.

Agnew, J., and Coleman, M.C. (eds) (2018) *Handbook on the Geographies of Power*, Edward Elgar Publishing.

Alaimo, S. (2014) 'Thinking as the stuff of the world', *O-Zone: A Journal of Object-Oriented Studies*, 1: 13–21.

Benessia, A., Funtowicz, S., Bradshaw, G., Ferri, F., Ráez-Luna, E.F., and Medina, C.P. (2012) 'Hybridizing sustainability: Towards a new praxis for the present human predicament', *Sustainability Science*, 7(1): 75–89.

Billig, M. (1997) 'The dialogic unconscious: Psychoanalysis, discursive psychology and the nature of repression', *British Journal of Social Psychology*, 36(2): 139–59.

Block, T., Goeminne, G., and Van Poeck, K. (2018) 'Balancing the urgency and wickedness of sustainability challenges: Three maxims for post-normal education', *Environmental Education Research*, 24(9): 1424–39.

Boiling, C. (2020) 'Is time part of terroir?', interview with A. Kirchner and K. Peter, International Wine Challenge, Available from: https://www.internationalwinechallenge.com/Canopy-Articles/is-time-part-of-terroir.html

Bucchi, M., and Trench, B. (2021) 'Introduction: Science communication as the social conversation around science', in B. Trench and M. Bucchi (eds) *Routledge Handbook of Public Communication of Science and Technology* (3rd edn), Routledge, pp 1–13.

Canfield, K.N., Menezes, S., Matsuda, S.B., Moore, A., Mosley Austin, A.N., Dewsbury, B.M. et al (2020) 'Science communication demands a critical approach that centers inclusion, equity, and intersectionality', *Frontiers in Communication*, 5(2).

Cerillo, C. (2024) 'The wine industry: Years of exploitation and human trafficking', The University of Alabama at Birmingham Institute for Human Rights Blog, Available from: https://sites.uab.edu/humanrights/2024/03/27/the-wine-industry-years-of-exploitation-and-human-trafficking/

Cooke, S.J., Gallagher, A.J., Sopinka, N.M., Nguyen, V.M., and Skubel, R.A. (2017) 'Considerations for effective science communication', *FACETS*, 2(1): 233–48.

Crang, M., and Thrift, N.J. (eds) (2000) *Thinking Space*, Routledge.

Cresswell, T. (2015) *Place: An Introduction* (2nd edn), J. Wiley & Sons.

Dagueneau, D. (2018) 'Excerpt from "Man and terroir"', in T. Patterson and J. Buechsenstein (eds) *Wine and Place: A Terroir Reader*, University of California Press, pp 16–17.

Davies, S.R., and Horst, M. (2016) *Science Communication: Culture, Identity and Citizenship*, Palgrave Macmillan UK.

Dawson, E. (2014) 'Reframing science communication: Engaging underrepresented audiences', *Public Understanding of Science*, 23(1): 17–27.

Duan, R., Huang, B., and Orthia, L.A. (2023) 'Shen Kua's Meng Hsi Pi T'an (c 1095 CE): China's first notebook encyclopaedia as a science communication text', in E. Rasekoala (ed) *Race and Sociocultural Inclusion in Science Communication: Innovation, Decolonisation, and Transformation*, Bristol University Press, pp 209–21.

Dürr, H.-P., Dahm, D., and zur Lippe, R. (2005) 'Potsdam Manifesto 2005: "We have to learn to think in a new way"', Federal Ministry of Education and Research of Germany, Available from: http://www.gcn.de/download/manifesto_en.pdf

Eisner, E.W. (2008) 'Persistent tensions in arts-based research', in M. Cahnmann-Taylor and R. Siegesmund (eds) *Arts-Based Research in Education: Foundations for Practice*, Routledge, pp 16–27.

Evans, H.J., and Achiam, M. (2021) 'Sustainability in out-of-school science education: Identifying the unique potentials', *Environmental Education Research*, 27(8): 1192–1213.

Felt, U. (2024) 'Time and temporalities in academic research', in U. Felt and A. Irwin (eds) *Elgar Encyclopedia of Science and Technology Studies*, Edward Elgar Publishing, pp 318–27.

Funtowicz, S.O., and Ravetz, J.R. (1993) 'Science for the post-normal age' *Futures*, 25(7): 739–55.

Gilbert, J.A., Van Der Lelie, D., and Zarraonaindia, I. (2014) 'Microbial terroir for wine grapes', *Proceedings of the National Academy of Sciences*, 111(1): 5–6.

Gregersen, E. (nd) 'Osmosis', *Encyclopaedia Britannica* [online], Available from: https://www.britannica.com/science/osmosis

Guattari, F. (1995) *Chaosmosis: An Ethico-aesthetic Paradigm*, translated by P. Bains and J. Pefanis, Indiana University Press.

Halpern, M.K., and Rogers, H.S. (2021) 'Art–science collaborations, complexities and challenges', in B. Trench and M. Bucchi (eds) *Routledge Handbook of Public Communication of Science and Technology* (3rd edn), Routledge, pp 214–37.

Haraway, D. (2008) *When Species Meet*, University of Minnesota Press.
Held, M. (2001) 'Sustainable development from a temporal perspective', *Time & Society*, 10(2–3): 351–66.
Howland, P.J. (2025) 'Times and temporalities', in P.J. Howland (ed) *Time and Alcohol*, Routledge, pp 6–25.
Horst, M., and Michael M. (2011) 'On the shoulders of idiots: Re-thinking science communication as "event"', *Science as Culture*, 20(3): 283–306.
Ingold, T. (2021) *Being Alive: Essays on Movement, Knowledge and Description* (1st edn), Routledge.
Jasanoff, S. (2005) *Designs on Nature: Science and Democracy in Europe and the United States*, Princeton University Press.
Jayer, H. (2018) 'Excerpt from "I have a dream"', in T. Patterson and J. Buechsenstein (eds) *Wine and Place: A Terroir Reader*, University of California Press, pp 13–14.
Johnson, J. (1988) 'Mixing humans and nonhumans together: The sociology of a door-closer', *Social Problems*, 35(3): 298–310.
Johnsson, M.C. (2015) 'Terroir and timespace: Body rhythms in winemaking', in B. Green and N. Hopwood (eds) *The Body in Professional Practice, Learning and Education: Body/Practice*, Springer International Publishing, pp 71–88.
Kankaria, S., Chakraborty, A., and Manna, A. (2023) 'Making knowledge visible: Artisans, craftsmen, printmakers, and the knowledge sharing practices of 19th-century Bengal', in E. Rasekoala (ed) *Race and Sociocultural Inclusion in Science Communication: Innovation, Decolonisation, and Transformation*, Bristol University Press, pp 222–38.
Kauffman, J. (2009) 'Advancing sustainability science: Report on the International Conference on Sustainability Science (ICSS) 2009', *Sustainability Science*, 4(2): 233–42.
Kramer, M. (2018) 'Excerpt from "The notion of terroir"', in T. Patterson and J. Buechsenstein (eds) *Wine and Place: A Terroir Reader*, University of California Press, pp 8–13.
Lang, D.J., Wiek, A., Bergmann, M., Stauffacher, M., Martens, P., Moll, P. et al (2012) 'Transdisciplinary research in sustainability science: Practice, principles, and challenges', *Sustainability Science*, 7(1): 25–43.
Latour, B. (2004) 'Why has critique run out of steam? From matters of fact to matters of concern', *Critical Inquiry*, 30(2): 225–48.
Latour, B., and Weibel, P. (eds) (2020) *Critical Zones: The Science and Politics of Landing on Earth*, MIT Press.
Leavy, P. (2020) *Method Meets Art: Arts-Based Research Practice* (3rd edn), Guilford Press.
Lewenstein, B. (1995) 'From fax to facts: Communication in the cold fusion saga', *Social Studies of Science*, 25(3): 403–36.
Lewenstein, B. (2007) 'Why should we care about science books?', *Journal of Science Communication*, 6(1).

Massey, D.B. (2012) *For Space*, Sage.

McNeill, J.R., and Engelke, P. (2014) *The Great Acceleration: An Environmental History of the Anthropocene since 1945*, Harvard University Press.

Morton, T. (2010) *The Ecological Thought*, Harvard University Press.

Myers, N. (2020) 'Becoming sensor in sentient worlds: A more-than-natural history of a black oak Savannah', in G.A. Bakke and M. Peterson (eds) *Between Matter and Method: Encounters in Anthropology and Art*, Routledge, pp 73–94.

Orthia, L.A. (2020) 'Strategies for including communication of non-Western and Indigenous knowledges in science communication histories', *Journal of Science Communication*, 19(2).

Paxson, H. (2010) 'Locating value in artisan cheese: Reverse engineering terroir for New-World landscapes', *American Anthropologist*, 112(3): 444–57.

Phillips, C. (2020) 'Telling times: More-than-human temporalities in beekeeping', *Geoforum*, 108: 315–24.

Rasekoala, E. (ed) (2023) *Race and Sociocultural Inclusion in Science Communication: Innovation, Decolonisation, and Transformation*, Bristol University Press.

Rigobello, A., and Evans, J.D. (2024) 'Design terroir: An eco-social, relational, bioregional approach to design', in C. Gray, E. Ciliotta Chehade, P. Hekkert, L. Forlano, P. Ciuccarelli, and P. Lloyd (eds) *Proceedings in the Design Research Society Library*, 23–8 June 2024, Boston, USA, Available from: https://doi.org/10.21606/drs.2024.269

Sedgwick, E.K. (1997) 'Paranoid reading and reparative reading; Or, you're so paranoid, you probably think this introduction is about you EVE KOSOFSKY SEDGWICK', in E.K. Sedgwick (ed) *Novel Gazing*, Duke University Press, pp 1–38.

Seguin, G. (1986) '"Terroirs" and pedology of wine growing', *Experientia*, 42: 861–73.

Sell, W. (2018) 'The war on terroir: Biology as (unstable) space in Jeff VanderMeer's Southern Reach trilogy', *Antae*, 5(1): 86–100.

Shome, R. (2003). 'Space matters: The power and practice of space', *Communication Theory*, 13(1): 39–56.

Sloterdijk, P. (2008) 'Foam city', *Distinktion: Journal of Social Theory*, 9(1): 47–59.

Taylor, S.S., and Ladkin, D. (2009) 'Understanding arts-based methods in managerial development', *Academy of Management Learning & Education*, 8(1): 55–69.

Teil, G. (2012) 'No such thing as terroir? Objectivities and the regimes of existence of objects', *Science, Technology, & Human Values*, 37(5): 478–505.

Tsing, A. (2010) 'Arts of inclusion, or how to love a mushroom', *Mānoa*, 22(2): 194–203.

van Leeuwen, C. (2009) 'Soils and terroir expression in wines', in E.R. Landa and C. Feller (eds) *Soil and Culture*, Springer Netherlands, pp 453–65.

Verlie, B. (2022) *Learning to Live with Climate Change: From Anxiety to Transformation*, Routledge.

Vostal, F., Benda, L., and Virtová, T. (2019) 'Against reductionism: On the complexity of scientific temporality', *Time & Society*, 28(2): 783–803.

Wynne, B. (1992) 'Misunderstood misunderstanding: Social identities and public uptake of science', *Public Understanding of Science*, 1(3): 281–304.

Ziebritzki, J. (2020) 'Sensorium of the earthbound', in B. Latour and P. Weibel (eds) *Critical Zones: The Science and Politics of Landing on Earth*, MIT Press, pp 260–3.

3

We Need to Imagine Sustainable Futures: Breaking with Modernity

Marianne Achiam

The Western world's failure to address looming global challenges is a failure of the imagination. Western societies remain captivated by modernity's materialistic-mechanistic worldview, constantly confronted with its logic in mass media, the education system, and societal discourse (Dewandre, 2018; Læssøe, 2020; Holten, 2024). This intellectual entrenchment in systems of capitalism, consumerism, and individualism hinders our ability to create solutions to our pressing problems. And those visions of the future we do produce tend to either be cynically dystopian or rely heavily on technology coming to our rescue (Haraway, 2016).

In this chapter, I explore the body of imaginations and futurisms generated by artists, scholars, and practitioners who work outside, sometimes in opposition to, the trappings and structures of Western modernity. As I hope to show, these imaginations offer rich, emotionally complex, hopeful, and creative ways of thinking about how to live within the planetary boundaries by prioritizing the needs of all life forms. Imagination, in this perspective, is more than fantasy, escapism, or a pastime for the elite; it is a social practice and a form of work (Appadurai, 1996) that can ultimately free up the energy necessary for breaking with our current patterns of living, being, and using the land (Moser, 2019).

In the following sections, I first substantiate the claim that a poverty of imagination is at the root of the global challenges we face. I then explain how I went about engaging with a diversity of critical imaginations and futurisms. I elaborate three themes that resonated with me in this engagement and, along the way, share the lived experiences, reflections, and imaginations this engagement prompted in me, in the form of brief autoethnographic vignettes (compare Green et al, 2021). I conclude with remarks on the implications of these ideas for science communication *as* sustainability.

What is imagination?

It is beyond the scope of this chapter to give an in-depth exposition of the uses and application of the term 'imagination' (for such an exposition, see, for example, Liao and Gendler, 2020). For the purposes of this text, imagination may be understood as the active process by which humans engage with the world by evoking or representing something that does not exist in that world or exists only as a possibility (Achiam, 2016). Imagination involves two seemingly opposing functions: the recreative imagination, which places oneself within the actual world from the perspective of one's past, future, or hypothetical self, or even another person; and the creative imagination, which can depart entirely from the actual world, generating concepts or solutions that may transcend reality (Currie, 2002). These components enable us to engage with and come to know reality but also – potentially – change it (Moore and Milkoreit, 2020).

Imagination plays a vital role in a variety of human activities and has been explored from multiple perspectives (Liao and Gendler, 2020). For instance, philosopher Kendall Walton (1993) describes how art generates prescriptions for imaginings by functioning as props in a game of make-believe. Walton's idea frames the aesthetic experience as a dynamic process where real-world artworks serve as springboards for imaginative activities, leading to fictional truths and scenarios that transcend the immediate physical reality of the artwork. This makes the experience of art an interplay between perceiving the actual artwork and engaging in make-believe that enriches and expands our experience beyond its tangible aspects (Wiltsher and Meskin, 2016).

This interaction between exploring the actual world and generating novel, sometimes surprising ideas is also a key part of science (Levy and Godfrey-Smith, 2020). Acts of imagination that connect the observable with the conceptual are at the heart of scientific reasoning (French, 2020). Thought experiments, analogies, models, and imaginary scenarios abound across scientific disciplines (Hadzigeorgiou, 2016; Salis and Frigg, 2020), clarifying points and sometimes dramatizing the discoveries or explanations they give rise to (Moue et al, 2006). Despite this, scientists rarely discuss the importance of imagination explicitly, and few, if any, Western education systems offer training in the use of imagination in scientific practice (Gough, 2024).

The poverty of the imagination

As the previous sections set out, the ideas and knowledge we encounter in our everyday lives shape our capacity to envision the future. This connection is particularly evident when we examine current sustainability strategies. Geoengineering technologies, such as those that capture CO_2 or increase atmospheric albedo, exemplify the 'technofixes' symptomatic of Western consumerism and perpetuate the idea of anthropocentric mastery of the world (Morozov, 2013; Koch, 2021). The creation of carbon markets shifts the burden of reducing emissions to the Global South, perpetuating colonialist and extractivist ideologies. Finally, the apparent inevitability of petro-capitalism, with its exploitation of the environment and its inhabitants, fosters a sense of defeat and cynicism, leading many to envision only dystopian futures (Haraway, 2016; Oziewicz, 2022). Immersed in the worldview of modernity, we struggle – individually and collectively – to conceptualize postmodern understandings of our place in nature (Riedy, 2022; Gough, 2024) and imagine desirable and sustainable futures (Haiven, 2014). This struggle is what a number of authors and scholars describe as the poverty of the imagination.

How can we move beyond this poverty? Modernity and the Western scientific worldview have historically placed significant domains of human experience, thought, and insight outside the domain of legitimate knowledge (Howitt and Suchet-Pearson, 2003). It is now past time to consider those domains. Accordingly, I draw on the growing body of imaginations that are being generated and articulated by feminist, Queer, and anti-colonialist scholarship, Indigenous scholarship and practice, Afrofuturisms, Africanfuturism, Gulf futurism, desi-futurism, Asian futurism, and futurism in Latin America. These perspectives offer a variety of creative and critical approaches to imagining the future. They cannot, and should not, be homogenized, but what they do have in common is that they collectively de-centre Western modernity as the sole founder of futurity (O'Connell, 2019). Importantly, as I hope to demonstrate, these perspectives challenge the reductionist view of reality as mere facts and figures by respecting the relationships and interconnections that constitute the complex web of life (compare Wildcat, 2009). In this, they allow us to understand and 'see' global problems such as climate change, environmental degradation, and biodiversity loss which are otherwise difficult to observe or experience in their entirety (Milkoreit, 2016).

Imaginations of the future

To the majority of the Earth's population, the sustainability problems we face are not new. Rather, they represent the continuation of the modernist practices of colonization, exploitation, and extractivism that have been at work for the last 500 years – the effects of which are just now reverberating back into the Western nations that introduced them (Davis and Todd, 2017). Accordingly, the hardships many people in Western societies are fearing are ones that Black, Indigenous, and People of Colour (BIPOC) as well as other marginalized groups have lived through already, due to different forms of injustice (Callison, 2014). In effect, these communities have already faced the end of a world (Mitchell and Chaudhury, 2020) and have developed tools of resilience and resourcefulness to endure (Samatar, 2017). First Nations (Cree-Métis) filmmaker Danis Goulet explains:

> I found that idea when I did research into Indigenous futurism – this idea that we've already survived the apocalypse – really interesting because I think it gives Indigenous artists a different way of coming at the apocalypse. In a colonial mindset, you're fearing the invasion, whether it's the alien invasion or the impending apocalypse or whatever, doom is going to happen. But from an Indigenous perspective, if that doom has already happened, then you might think that we are better equipped to deal with that when it happens. And in terms of climate catastrophe, [the question is] what can we look at from Indigenous worldviews and values that could help us in the situation that we all find ourselves in, whether we're Indigenous or not? (Goulet, about 00.17.55 in Molinsky, 2022)

This kind of fluid and decolonial imagination is urgently needed to counteract modernity's 'fixed tropes of progress and innovation' that keep us from imagining and creating worlds in which all humans can thrive (Benjamin, 2024, p 114).

Methodology

The content I present in the following sections is generated through a practice inspired by Leavy (2020), in which arts-based methods and more traditional, qualitative research methods inform each other. I situate this practice within the overarching critical-emancipatory paradigm, meaning that I see the object of study – imaginations of the future – as being shaped by ideology and hegemony as outlined in the preceding sections, and I see my work as a potential albeit modest contribution to reform and change (compare Miller, 2005). On the one hand, I draw on the qualitative method of thematic analysis (for example, Terry et al, 2017) to identify ideas and insights that cut across the critical feminist, Queer, and BIPOC scholarship and practice I engage with. On the other hand, I aim for neither generalizability nor transferability in my analysis, but rather *resonance* (Leavy, 2020). By resonance, I mean my experience of becoming emotionally, bodily, and cognitively attuned to what I was engaging with – 'those moments when something crackles or reverberates or comes alive' (Felski, 2020, p 2).

It follows that the themes I share in the following emerged from an organic and flexible process in which I engaged with the work of artists, writers, and activists who imagine futures as well as that of scholars who interrogate, philosophize, and theorize those imaginations (compare Yaszek, 2006). My analysis is situated at the intersection of the work I engaged with, the theoretical and conceptual frameworks this work adhered to, my disciplinary knowledge (that is, of science communication, sustainability, and the role of imagination), and my research skills and experience (compare Terry et al, 2017). My analysis and synthesis are thus not simply waiting in the data to be found but are instead subjective and reflexive.

As a white cishet female academic, a graduate of a Western science (biology), and a citizen of a colonial nation (Denmark), I am aware that my analysis and synthesis run the risk of making unconscious assumptions that perpetuate race, class, ontology, gender, and settler status as power structures, even as I attempt to confront those assumptions (compare Mitchell and Chaudhury, 2020). In the final remarks of this text, I attempt to examine how my 'thoughts and desires reflect and reproduce a dominant imagination that values some lives over others' (Benjamin, 2024, p 64) and reflect on the limitations of this approach.

Imagining sustainable futures

In my engagement with the practices and scholarship of feminist, Queer, and BIPOC futurism, three themes resonated with me, namely temporality, subjectivity, and geography – or time, body, and space (compare Howitt and Suchet-Pearson, 2003; D'Souza, 2019; Mitchell and Chaudhury, 2020). I discuss these themes in the following sections, describing how they play out across the critical literature I engaged with and making suggestions for their practical implications.

Temporality, or time

Modernity represents time as linear and homogenous (Held, 2001), and the English language embeds this linearity by viewing time as categorically distinct in past, present, or future tenses (Howitt and Suchet-Pearson, 2003). This linear perspective shapes societal narratives and influences how we think about various issues, including sustainability. For instance, popular climate catastrophe narratives often focus on the *future* destabilization of Western privilege and ignore the *ongoing* climate and environmental injustice (Evans, 2017). At the same time, Western societal discourse about sustainability problems presents time as scarce or running out, locking people into the crisis-response model and leaving little opportunity for communities and individuals to plan for their futures (Eichelbaum et al, 2001; Phillips, 2022).

In contrast to the Western linear conception of time, Potawatomi scholar and activist Kyle P. Whyte discusses how 'spiraling time is an important topic of discussion when Indigenous persons compare their conceptions of temporality across different cultures' (Whyte, 2018, p 229). He describes the Anishinaabe perspective on time as intergenerational, in which people live alongside future and past relatives simultaneously as they spiral through time. He thinks of this intergenerational perspective as a way to ask how past and future generations would understand the critical situations we are in today, in order to respond appropriately. In Maori understandings of temporality, as well, the spiral is an important metaphor, representing pre-existence and potentiality, offering many layers of meaning to time. The present and the future thus become indistinguishable from the past (Stewart-Harawira, 2005). Long-deceased ancestors don't become more and more distant but continue to be a part of current groups and to inhabit the places they lived and where they left tangible and intangible remains (Steeves, 2015). Desi-futurism similarly rejects the idea of linear time, using the Hindi word *kal* for tomorrow as well as yesterday; the past as well as the future (D'Souza, 2019).

Critiques of linear, progressive time have also been articulated forcefully within Black and Afrofuturist scholarship, where temporal models are directly linked to the histories and ongoing effects of colonialism and racial capitalism. Rasheedah Phillips (2022), a Black artist, lawyer, and scholar, interrogates the linear, mechanical conception of time that underpins Western modernity, demonstrating how it produces and legitimizes hierarchies that have historically marginalized Black people and communities. She argues that Afrofuturist perspectives offer alternative temporal frameworks that displace linear progress with a social conception of time understood as affective, layered, durational, and context dependent. These temporalities are rooted in everyday practices and specific places, and they challenge the assumption that time unfolds in a universal, forward-moving trajectory.

Phillips describes how rejecting dominant temporal assumptions such as time as scarcity, productivity, or monetary value opens up alternative modes of survival and world making within systems structured by those assumptions:

> If I don't assume that time runs linearly or that I'm running out of time, or that time is equated with money, or that my value as a human being is embedded in time and money, then that just opens up so much for me in terms of what is possible, and how I'm going to construct the tools that I need to survive in a world that does believe that when I don't believe that. (Phillips in McEwen and Garcia, 2023, p 22)

Together with Camae Ayewa, Phillips developed Black Quantum Futurism, a practice that reimagines lived reality by actively manipulating space and time to envision and enact desired futures. Drawing on both quantum physics and Black African cultural traditions, Black Quantum Futurism conceptualizes time and space as dynamic, relational, and co-constitutive of identity. Rather than treating the past and future as fixed or distant, this framework understands them as continually shaping present experience and possibility (Carter, 2022).

Collectively, then, Black and other BIPOC futurist traditions articulate what Mitchell and Chaudhury (2020) describe as forms of *temporal sovereignty*: refusals of externally imposed timelines that frame progress, crisis, and sustainability through Eurocentric developmental logics. From this perspective, sustainability challenges are not abstract futures 'ahead of us' but conditions that have long been lived and negotiated in the aftermath of colonial violence, dispossession, and environmental extraction (Whyte, 2018). These temporal frameworks therefore contest modernist claims that humanity is inherently constrained by short time horizons and incapable of addressing long-term sustainability problems, showing instead how alternative temporalities have already supported endurance, adaptation, and world rebuilding under conditions of profound structural harm.

Autoethnographic vignette: Seeing the future

I have lost count of the number of climate exhibitions I have visited.[1] The graphs, the figures, the timelines stretching towards 2030, 2050 ... the looming deadlines for catastrophe. 'Fossil fuels account for 80 per cent of global energy. The planet's temperature has risen by one degree since the 1800s.' The screens glow sterile blue while the numbers stack into inevitability. There are silhouettes of people, but no names, no stories, no presence. I read the text, nod along. I understand, but I feel nothing. Then, I attended a presentation by the author Rob Hopkins, founder of the Transition Movement. He invited us to step into a different kind of temporality, not one that raced towards collapse but one that spiralled towards possibility. He asked us to travel with him to the future. The lights dimmed, and voices from the future filled the space. People – real people – spoke to us across time, looking back at 2023.

> A teacher described the school she had helped transform. Its walls were covered in vines, and its curriculum was centred on care. A former parking garage operator walked us through his mushroom farm, thriving where cars had once idled. A community organizer told us how they had closed their streets to traffic and turned them into gardens. These futures felt tangible – not abstract blueprints or policy projections but lived experiences, thick with texture and emotion. Then came the reveal: Everything you just heard is real. These initiatives already exist. I felt something shift inside me. For the first time in a long time, I was not just informed about sustainability – I was moved by it. I thought about temporal sovereignties: the refusal to be bound by crisis time, the insistence that alternative worlds were not distant dreams but ongoing, unfinished realities. I stepped out of the keynote thinking: This is how we should be telling the story of our future.

Subjectivity, or bodies

Modernity separates humans from other living beings. Indeed, the conventional Western attitude is to assume fundamental differences between people and animals (Scott, 2011; Suchet-Pearson et al, 2013). Potawatomi scholar Robin W. Kimmerer explains how the English language reinforces this separation by either reducing a non-human to an *it* or gendering the non-human inappropriately as a *she* or *he*. 'The arrogance of English is that the only way to be animate, to be worthy of respect and moral concern, is to be a human' (Kimmerer, 2020, pp 56–7). This disregard is reflected in modernist humanity's demand to control the world, which hinders animals, plants, and other life forms from continuing their sustainable living (Benessia et al, 2012).

In contrast, BIPOC futurisms often feature subjects that are not only of mixed heritage but also part of non-human communities that include animals and land (Mitchell and Chaudhury, 2020). For instance, much Indigenous knowledge does not distinguish between humans and nature. In Cree, anthropologist Colin Scott (2011) explains, there is no word that corresponds to the Western term 'nature'; rather, there is a word which includes human as well as animal persons. In this worldview, 'humans, animals, spirits, and several geophysical agents are perceived to have qualities of personhood' (p 179). Further, Cree scholarship assumes that there are shared connections among people, animals, and other entities (Scott, 2011). Many Aboriginal people, as well, have relationships with animals based on an underlying understanding of the common heritage and kinship between animals and humans (Howitt and Suchet-Pearson, 2003). Yolŋu ontology, Suchet-Pearson et al (2013) explain, considers humans and non-humans to only exist through their relationships. The wellbeing of the non-human is thus entangled with the wellbeing of the human, and they sustain each other in a reciprocal relationship.

Critical feminist scholar Donna Haraway frames a similar discussion in terms of 'making kin' (Haraway, 2016). Kin, in Haraway's view, includes not just genealogical and biogenetic family but also other-than-human beings; making kin thus redefines to whom we are actually responsible. Haraway uses the term 'tentacular' to describe the complex and recursive interconnections of multispecies assemblages, effectively rejecting the modernist notion of the (living) world as being composed of well-delineated, discrete units (see Chapters 5 and 8 for more on making kin with multispecies assemblages).

Considerations of subjectivity in relation to sustainability must also include a reckoning with the so-called 'cult of John', the evocation in Western popular culture of a rugged Caucasian male protagonist dressed in safari gear (or more recently a lab coat), tasked with saving the world (Mitchell and Chaudhury, 2020). This trope epitomizes the cis-hetero-patriarchal norms and forms of oppression that are considered timeless and inevitable in much Western culture (Whyte, 2018) and that reinforce binaries such as male/female, human/animal, and domestic/wild (Howitt and Suchet-Pearson, 2003). However, as Queer Indigenous futurist Pınar Sinopoulos-Lloyd (nd) explains, when we only acknowledge the world in terms of binaries, the *liminal walkers* (Queer, trans, and other beings) continue to be ignored and erased despite the unique role they have to play in 'reinvigorating the creative solutions and alternative lifeways necessary in this time of disruption and collapse' (Raditz, 2024, p 164).

More generally, Queer resilience seeks to challenge monocultural (for instance, cis-hetero-patriarchal) ways of life by drawing on the ecological principle that diverse ecosystems are stronger and more resilient (Raditz, 2024). Researcher and speculative designer Jason Tester emphasizes that this diversity and fluidity – embracing the richness of everyone's identities and rejecting simplistic binaries – is central to the mindsets and practices of queerness. Queering the future, in his vision, involves finding comfort, and even pleasure and joy, outside categorization. The future is defined, Tester affirms, by the ability to reframe, repurpose, and 'hustle between' different worlds (in Larsen, 2021).

In summary, the critical futurisms I engaged with collectively rebuff the limits imposed on humans and non-humans by modernity: patriarchy, colonialism, racism, speciesism, and cis-heteronormativity (Medak-Saltzman,

2017; Bigelow, 2021). The joyful rejection of binary categories advocated by Queer scholars and practitioners allows for a cultivation of rich solidarities across communities and even species (Mitchell and Chaudhury, 2020) that may enable us to 'see and feel beyond the quagmire of the present' (Muñoz, 2019). Just as importantly, these potential solidarities necessarily foreground our responsibility and accountability to the world and all its non-human inhabitants.

Autoethnographic vignette: Making kin, making futures

I used to believe in the 'natural' order of things: Humans above animals; science above sentiment.[2] In my biology studies, life was something to be understood by breaking it apart. We dissected animals, prepared tissue samples, studied nerves. A classmate caused a ruckus when he refused, objecting on moral grounds. I thought he was naïve: how could one be a scientist without studying animals, and didn't studying mean sacrificing? I ate meat without question. Animals ate other animals. I had been taught that their suffering was different – less complex, less real. It was convenient to believe this. Then, years later, I read a paper about slaughterhouse workers. It described how they coped with the daily act of killing: some dissociated, some joked, but most reframed reality. Cows weren't living beings; they were food-in-waiting. Their eyes didn't hold recognition, and their bodies were just objects in motion. It was the only way to endure the work. Around the same time, I got a puppy. She was small, warm, trusting. She followed me from room to room, her eyes locked onto mine, and suddenly, the distance I had once believed in – the certainty that humans were separate, superior – collapsed. What if science had gotten it wrong? What if the world was not neatly divided into categories, but tangled and entangled, in Donna Haraway's words, tentacular? In another timeline, sustainability communication would start here. Not with sterile graphs, not with crisis countdowns, but with stories of kinship. What if we didn't just communicate sustainability as data, but as relations to be nurtured? What if we designed exhibitions where visitors stepped into the world as a river, a bird, a forest? What if sustainability education didn't reinforce human dominance but invited us to feel our entanglements, our responsibilities? I no longer believe in the natural order of things. I believe in making kin, and in making futures.

Geography, or space

Modernity portrays itself as transcending its earthly roots, liberated from the constraints of nature (Bonneuil and Fressoz, 2017). The discovery and exploitation of fossil fuels reinforced this sense of detachment: it appeared as though nature had bestowed an inexhaustible reservoir of energy on humanity, freeing us from the limits of land-based resources (Say, 1840). This perception of dominion over nature paved the way for the imposition of universalist (Eurocentric) ideals by colonizers, who often ignored the distinct realities of the landscapes they encountered, instead reshaping them to fit an idealized, homogenized vision reminiscent of their homelands (Davis and Todd, 2017). The drive for uniformity extended to the discipline of geography, exemplified by the designation of Greenwich, England, as the prime meridian. This decision not only standardized global time and space but also symbolized the imposition of European spatial frameworks on a diverse world (D'Souza, 2019).

The infliction of modernity's homogenizing ideals on the diverse landscapes of the Global South has had severe repercussions. From collapsed ecosystems to enduring poverty, the colonial legacy has left lasting scars on these regions (Samatar, 2017; Tsing, 2017). Further, the redistribution of resources under colonial rule entrenched a global hierarchy, wherein wealth and access became concentrated in the Global North, perpetuating a cycle of dependency and disparity (Pirker and Rahn, 2020; Machado de Oliveira, 2021). Thus, while the Global North is only beginning to reckon with the environmental fallout of industrial modernity, the Global South has long been its unwilling front line.

In the African continent, the lingering effects of modernity have prompted counter-conceptualizations of the future that reimagine representations of Africa and Africans (Pirker and Rahn, 2020). Nigerian American science-fiction writer Nnedi Okorafor coined the term Africanfuturism to describe a way of thinking that is rooted in African culture, history, mythology and point of view, and is concerned with 'what is and can/will be' (Okorafor, 2019).

Much Africanfuturism focuses on Africans' resilience and resourcefulness in the face of climate change, as explained by Nigerian speculative-fiction writer Chinelo Onwualu:

> When we are talking about cli-fi a lot of the writers that I've encountered are not approaching it as the sort of subgenre of a subgenre. It's kind of built into how many of us understand our future. We are seeing the effects of climate change in ways that are much more immediate. (Onwualu, about 00.25.15, in Molinsky, 2024)

Africanfuturism envisions human existence in entanglement with its living and non-living world; thus, planetary habitability does not privilege or prioritize the human over non-human forms of existence (Ogwu, 2022). This reimagining of the ongoing environmental crises as entanglements of humans and non-humans with the world helps to better understand the complexity and urgency of the problems in the context of specific landscapes and ecosystems (Young, 2023).

The modernist division of mind and thought from land also runs counter to many Indigenous ways of seeing the world (Davis and Todd, 2017). The Yolŋu ontology of becoming, for instance, sees Country as 'humans as well as waters, seas and all that is tangible and non-tangible' (Suchet-Pearson et al, 2013, p 186). In this ontology, the nature of being is relational, and living and non-living beings *become* through relations in a domain of entanglement (Ingold, 2006). In Inuit knowledge, as well, history and context matter – and they vary between villages, regions, and countries. Social issues and mental wellbeing are intimately connected to the environment (Callison, 2014), as explained by Patricia Cochran, a former International Chair of the Inuit Circumpolar Council:

> All things are connected, and so to take one piece of a problem and not connect it to the rest of the world and the environment around? It just logically makes no sense. How can we talk about changes in weather without talking about changes in vegetation or the air or the people or the animals, as all of those things are part of a natural mix. All things are connected in our universe. (Cochran, in Callison, 2014, p 46)

There are numerous examples of how the imposition of nation-state structures and borders bind semi-nomadic societies inside given territories, hindering their access to the living and non-living assemblages (rocks, water, plants, animals) they are entwined with (Callison, 2014; Davis and Todd, 2017). Critical futurisms reject the imposition of such borders. Desi-futurism, for instance, imagines a South Asian homeland that existed prior to the imposition of borders and diasporas, allowing for the crystallization of a shared identity among desis (D'Souza, 2019).

In summary, the critical futurisms I engaged with challenge modernity's imposed spatial frameworks, which have reshaped landscapes through Eurocentric ideals, often at great ecological and social cost. While modernity seeks to dominate and transcend the land, Indigenous, Africanfuturist, and desi-futurist perspectives emphasize relationality and deep historical continuity between people and geography. These perspectives assert that the spatial, the natural, and the social co-construct each other (Howitt and Suchet-Pearson, 2003) and frame sustainability as an ongoing, situated practice rather than an abstract, future goal. By focusing on resilience, adaptation, and spatial justice, they reveal that sustainability is inseparable from questions of access, autonomy, and the right to shape one's own environment along with the rhythms of a changing planet (Mitchell and Chaudhury, 2020).

Autoethnographic vignette: The memory of land

I used to love dioramas.[3] Their precise and complete miniature worlds felt like portals to places I might never visit. An Arctic ice field, a sunlit savanna, a dense rainforest – each a window on the world, perfectly arranged for my observation. But over time, I began to see the cracks in the glass. Dioramas weren't snapshots of reality but constructed narratives, which reinforced national pride, gendered exploration, and human dominance over nature. They framed landscapes as fixed, unchanging, waiting to be discovered, named, and categorized. I had never questioned why these windows never showed pipelines, fences, or the traces of those displaced from the land. Later, I visited the Forensic Architecture exhibition. One installation told a different story – of memory, violence, and reclamation. Through digital reconstructions, campesinos (small-scale farmers) in Colombia identified landscapes they had been dispossessed of: tracing where rivers once ran before drought, where forests had been cleared, where erosion had swallowed familiar markers. The earth itself had been altered, but it had not forgotten. Plants still tried to return where monocultures had stripped the soil bare. Water carved new pathways, remembering old ones. Seeds lay dormant, waiting. The land was not just a victim of extraction – it was a witness. It remembered. What if sustainability communication recognized land in this way: not as scenery or resource, but as testimony? What if, instead of graphs and maps, we listened to the ways land speaks? What futures could we imagine if landscapes were not backdrops but active participants in our survival? I think of the dioramas. I once believed they were windows on the world. But now I wonder – what stories do they refuse to tell?

Concluding remarks: Telling stories from the margins, from the centre

I began this text by questioning how to move beyond the poverty of the imagination, arguing that Western modernity limits us to futurisms of either catastrophe or technocratic salvation. As an alternative to those visions, I explored the relational, entangled visions of the future imagined by feminist, Queer, and BIPOC scholars and practitioners. Together, these imaginings reveal that sustainability is not just a scientific or technological challenge; it is a deeply political, cultural, and ontological question about who gets to imagine the future – and whose futures are made possible. The futurisms I explored here prompted me to reflect carefully about my own lived experiences with sustainability science communication, and to imagine new ones that embody science communication *as* sustainability. This prompts me to interrogate my own positionality (compare Raditz, 2024) by asking: who am I to draw on and utilize the critical voices of feminist, Queer, and BIPOC communities? What right do I have to represent their lives? Which stories do I tell, and which do I leave out?

Given my identity – white, cishet female, graduate of a Western science, citizen of a colonial nation – I recognize that I am always at risk of reproducing the power structures I aim to critique. For instance, my training in biology taught me to see the world through a lens of categorization, objectivity, and control – a lens that modernity has wielded since the Enlightenment to separate humans from nature, to define which bodies and lives are legible, and to determine who speaks for whom. Even as I attempt to dismantle these assumptions, I remain shaped by them.

Although I see myself as a feminist, I cannot claim to speak for Queer and BIPOC communities. Rather, my aim is to enquire how knowledge is produced, whose voices shape the dominant discourse, and how I can decentre myself in my own process of learning. Part of this enquiry is acknowledging that silence is not neutrality: to refuse to engage with Queer and BIPOC stories out of fear of misrepresentation is to uphold the status quo, allowing dominant narratives – of crisis, of inevitability, of white saviourism – to persist unchallenged. Accordingly, the question for me is not *whether* to engage, but *how*. How can I carry out this work while maintaining the richness and complexity of the material? How can I hold space for multiple ways of knowing without absorbing them into a Western academic framework? How can I remain accountable to the stories I tell – and to those I leave out?

This chapter is not an answer but an attempt. It is a recognition of my limitations, of the risks of appropriation, and of the necessity of discomfort in doing work that unsettles dominant structures. If science communication is to move towards more sustainable futures, then those of us who have benefitted from the exclusions of modernity must first ask: what does it mean to tell stories from the centre, about the margins? And how do we learn to listen, instead?

Implications for science communication as sustainability

Science communication *as* sustainability is not just about conveying facts – it is about shaping how people imagine, experience, and relate to possible futures. The perspectives I have explored in this text offer different ways of knowing, relating, and imagining. They reject linear, technocratic solutions and instead emphasize relationality, entanglement, and the multiplicity of possible futures. For science communication researchers and practitioners, the challenge is: how do we translate these insights into action? How do we move beyond dominant modes of crisis communication to approaches that resonate, inspire, and open up new possibilities? In the last part of this text, I share a series of guiding questions drawn from the ideas, reflections, and imaginations in this chapter. These questions are not meant as a checklist but as starting points for science communicators (be they researchers or practitioners) who are struggling with how to engage audiences differently. As with sustainability itself, the work of rethinking science communication *as* sustainability is ongoing, unfinished, and always in the making.

Making futures tangible:

- How can we move beyond presenting sustainability as a distant crisis towards something that is already unfolding?

- What storytelling techniques can help audiences experience rather than just learn about sustainability?
- How can we incorporate participatory or speculative methods – such as future visioning, immersive storytelling, or embodied experiences – into science communication?

Challenging human exceptionalism:

- How do we frame sustainability beyond human interests, acknowledging the agency of landscapes, animals, and ecosystems?
- What new metaphors or framings could disrupt the anthropocentric focus in sustainability narratives?
- How can we design multisensory or embodied experiences that help audiences feel their entanglement with the more-than-human world?

From crisis narratives to relational accountability:

- How can we move away from urgency-driven messages without minimizing the stakes?
- How can we communicate long-term responsibility rather than short-term emergency?
- What examples of intergenerational, community-led sustainability efforts could provide alternative models for communication?

Whose stories get told?

- How do we ensure that sustainability communication does not flatten diverse experiences into a single narrative?
- How do we engage with feminist, BIPOC, and Queer perspectives without extractive storytelling?
- How can communication be structured to centre listening, reciprocity, and co-creation rather than one-directional dissemination?

Notes

1. The vignette 'Seeing the future' refers to climate and sustainability exhibitions; these are briefly discussed in Evans and Achiam (2021). The keynote presentation by Rob Hopkins was entitled 'From what is to what if: Why the next 10 years needs to feel like a revolution of the imagination', delivered at the 2023 Ecsite conference in Valletta, Malta.
2. The vignette 'Making kin, making futures' refers to the paper 'Knowing cows: Transformative mobilizations of human and non-human bodies in an emotionography of the slaughterhouse' by McLoughlin (2019).
3. The vignette 'The memory of land' refers to the exhibition 'Forensic architecture: Witnesses', on display at Louisiana Museum of Modern Art in Humlebæk, Denmark in the period 20 May–23 October 2022.

References

Achiam, M. (2016) 'The role of the imagination in museum visits', *Nordisk Museologi*, 16(1): 89–100.

Appadurai, A. (1996) *Modernity at Large: Cultural Dimensions of Globalization*, University of Minnesota Press.

Benessia, A., Funtowicz, S., Bradshaw, G., Ferri, F., Ráez-Luna, E.F., and Medina, C. P. (2012) 'Hybridizing sustainability: Towards a new praxis for the present human predicament', *Sustainability Science*, 7(1): 75–89.

Benjamin, R. (2024) *Imagination: A Manifesto*, W.W. Norton & Company.

Bigelow, C. (2021) 'Queer futurism', Medium, Available from https://medium.com/prismnpen/queer-futurism-9b2f8ca1671d

Bonneuil, C., and Fressoz, J.-B. (2017) *The Shock of the Anthropocene: The Earth, History and Us* (D. Fernbach, trans), Verso. [Original work *L'événement Anthropocène: La Terre, l'histoire et nous* published in 2013.]

Callison, C. (2014) *How Climate Change Comes to Matter: The Communal Life of Facts*, Duke University Press.

Carter, B.W. (2022) *Afrofuturism and Digital Humanities: Show Me and I Will Engage Differently*, Routledge.

Currie, G. (2002) 'Projections and recreations', in G. Currie and I. Ravenscroft (eds) *Recreative Minds*, Oxford University Press, pp 7–23.

D'Souza, R.A. (2019) 'Imagining a desi future', *South Asian Popular Culture*, 17(1): 47–59.

Davis, H., and Todd, Z. (2017) 'On the importance of a date, or decolonizing the Anthropocene', *ACME: An International Journal for Critical Geographies*, 16(4): 761–80.

Dewandre, N. (2018) 'Political agents as relational selves: Rethinking EU politics and policy-making with Hannah Arendt', *Philosophy Today*, 62(2): 493–519.

Eichelbaum, T., Allan, J., Fleming, J., and Randerson, R. (2001) 'Report on genetic modification', Royal Commission on Genetic Modification.

Evans, H.J., and Achiam, M. (2021) 'Sustainability in out-of-school science education: Identifying the unique potentials', *Environmental Education Research*, 27(8): 1192–1213.

Evans, R. (2017) 'Fantastic futures? Cli-fi, climate justice, and Queer Futurity', *Resilience: A Journal of the Environmental Humanities*, 4(2–3): 94–110.

Felski, R. (2020) 'Resonance and education', *On Education: Journal for Research and Debate*, 3(9): 1–4.

French, S. (2020) 'Imagination in scientific practice', *European Journal for Philosophy of Science*, 10(3): 27.

Gough, N. (2024) 'Exercising the imagination: Ecofeminist science fictions as object-oriented thought experiments in education', *Gender and Education*, 36(4): 345–61.

Green, D., Levey, A., Evans, B., Lawson, W., and Marks, K. (2021) 'The arts of making-sense in uncertain times: Arts-based research and autoethnography', in H. Kara and S.-m. Khoo (eds) *Qualitative and Digital Research in Times of Crisis: Methods, Reflexivity and Ethics*, Bristol University Press, pp 59–77.

Hadzigeorgiou, Y. (2016) *Imaginative Science Education*, Cham: Springer.

Haiven, M. (2014) *Crises of Imagination, Crises of Power: Capitalism, Creativity and the Commons*, Bloomsbury Academic & Professional.

Haraway, D.J. (2016) *Staying with the Trouble: Making Kin in the Chthulucene*, Duke University Press.

Held, M. (2001) 'Sustainable development from a temporal perspective', *Time & Society*, 10(2–3): 351–66.

Holten, E. (2024) *Underskud: Om værdien af omsorg* [Deficit: On the value of care], Politikens Forlag.

Howitt, R., and Suchet-Pearson, S. (2003) 'Ontological pluralism in contested cultural landscapes', in K. Anderson, M. Domosh, S. Pile, and N. Thrift (eds) *Handbook of Cultural Geography*, SAGE Publications, pp 557–69.

Ingold, T. (2006) 'Rethinking the animate, re-animating thought', *Ethnos*, 71(1): 9–20.

Kimmerer, R.W. (2020) *Braiding Sweetgrass: Indigenous Wisdom, Scientific Knowledge and the Teachings of Plants*, Penguin Books.

Koch, N. (2021) 'Whose apocalypse? Biosphere 2 and the spectacle of settler science in the desert', *Geoforum*, 124: 36–45.

Læssøe, J. (2020) 'Dynamisk Samspilstænkning' [Dynamic interactional thinking], in J.A. Lysgaard and N.J. Jørgensen (eds) *Bæredygtighedens Pædagogik: Forskningsperspektiver og Eksempler fra Praksis* [The pedagogy of sustainability: Research perspectives and examples from practice], Frydenlund, pp 107–21.

Larsen, N. (2021) 'Queer futures', Medium, Available from https://medium.com/copenhagen-institute-for-futures-studies/queer-futures-with-jason-tester-43c4e65cdfd1

Leavy, P. (2020) *Method Meets Art: Arts-Based Research Practice* (3rd edn), Guilford Press.

Levy, A., and Godfrey-Smith, P. (2020) 'Introduction', in A. Levy and P. Godfrey-Smith (eds) *The Scientific Imagination*, Oxford University Press, pp 1–16.

Liao, S.-y., and Gendler, T. (2020) 'Imagination', in E.N. Zalta (ed) *The Stanford Encyclopedia of Philosophy (Summer 2020 Edition)*, Available from: https://plato.stanford.edu/archives/sum2020/entries/imagination/

Machado de Oliveira, V. (2021) *Hospicing Modernity: Facing Humanity's Wrongs and Implications for Social Activism*, North Atlantic Books.

McEwen, V.M., and Garcia, C. (2023) 'Holding the Future: In conversation with Rasheedah Phillips', *Journal of Architectural Education*, 77(1): 20–7.

McLoughlin, E. (2019) 'Knowing cows: Transformative mobilizations of human and non-human bodies in an emotionography of the slaughterhouse', *Gender, Work & Organization*, 26(3): 322–42.

Medak-Saltzman, D. (2017) 'Coming to you from the Indigenous Future: Native women, speculative film shorts, and the art of the possible', *Studies in American Indian Literatures*, 29(1): 139–71.

Milkoreit, M. (2016) 'The promise of climate fiction: Imagination, storytelling, and the politics of future', in P. Wapner and H. Elver (eds) *Reimagining Climate Change*, Routledge, pp 171–91.

Miller, K. (2005) *Communication Theories: Perspectives, Processes and Contexts*, McGraw Hill.

Mitchell, A., and Chaudhury, A. (2020) 'Worlding beyond "the" "end" of "the world": White apocalyptic visions and BIPOC futurisms', *International Relations*, 34(3): 309–332.

Molinsky, E. (2022) 'Indigenous futurisms' (no 211) [audio podcast episode], *Imaginary Worlds*, 10 November, Available from: https://www.imaginaryworldspodcast.org/episodes/indigenous-futurisms

Molinsky, E. (2024) 'African sci-fi looks to a future climate' (no 248) [audio podcast episode], *Imaginary Worlds*, 10 April, Available from: https://www.imaginaryworldspodcast.org/episodes/african-sci-fi-looks-to-a-future-climate

Moore, M.-L., and Milkoreit, M. (2020) 'Imagination and transformations to sustainable and just futures', *Elementa: Science of the Anthropocene*, 8(1): 1–17.

Morozov, E. (2013) *To Save Everything, Click Here: The Folly of Technological Solutionism*, Public Affairs.

Moser, S. (2019) 'Tasks of climate change communication in the context of societal transformation', in G. Feola, H. Geoghegan, and A. Arnall (eds) *Climate and Culture: Multidisciplinary Perspectives on a Warming World*, Cambridge University Press, pp 141–67.

Moue, A.S., Masavetas, K.A., and Karayianni, H. (2006) 'Tracing the development of thought experiments in the philosophy of natural sciences', *Journal for General Philosophy of Science*, 37(1): 61–75.

Muñoz, J.E. (2019) *Cruising Utopia, 10th Anniversary Edition*, New York University Press.

O'Connell, H.C. (2019) 'Science fiction and the Global South', in G. Canavan and E.C. Link (eds) *The Cambridge History of Science Fiction*, Cambridge University Press, pp 680–95.

Ogwu, E.O. (2022) 'Humanism made to the measure of the world: The case of Africanfuturism', [master's thesis], University of the Witwatersrand, Available from: https://www.proquest.com/openview/2405a94869eabb61d11e0d13e7cd6688/1?pq-origsite=gscholar&cbl=2026366&diss=y

Okorafor, N. (2019) 'Africanfuturism defined', Nnedi's Wahala Zone Blog, Available from: https://nnedi.blogspot.com/2019/10/africanfuturism-defined.html
Oziewicz, M. (2022) 'Planetarianism now: On anticipatory imagination, young people's literature, and hope for the planet', in M. Paulsen, J. Jagodzinski, and S.M. Hawke (eds) *Pedagogy in the Anthropocene. Re-wilding Education for a New Earth*, Palgrave Macmillan, pp 241–56.
Phillips, R. (2022) 'Race against time: Afrofuturism and our liberated housing futures', *Critical Analysis of Law: An International & Interdisciplinary Law Review*, 9(1): 16–34.
Pirker, E.U., and Rahn, J. (2020) 'Afrofuturist trajectories across time, space and media', *Critical Studies in Media Communication*, 37(4): 283–97.
Raditz, V. (2024), 'Visibilizing Queer resilience: Representational justice for the climate movement', in J.L. Rice, J. Long, and A. Levenda (eds) *Urban Climate Justice: Theory, Praxis, Resistance*, University of Georgia Press, pp 167.
Riedy, C. (2022) 'Discursive entrepreneurship: Ethical meaning-making as a transformative practice for sustainable futures', *Sustainability Science*, 17(2): 541–54.
Salis, F., and Frigg, R. (2020) 'Capturing the scientific imagination', in A. Levy and P. Godfrey-Smith (eds) *The Scientific Imagination*, Oxford University Press, pp 17–50.
Samatar, S. (2017) 'Toward a planetary history of Afrofuturism', *Research in African Literatures*, 48(4): 175–91.
Say, J.-B. (1840) *Cours complet d'economie politique practique* [Complete treatise on practical political economy], Guillaume.
Scott, C. (2011) 'Science for the West, myth for the rest? The case of James Bay Cree knowledge construction', in S. Harding (ed) *The Postcolonial Science and Technology Studies Reader*, Duke University Press, pp 175–97.
Sinopoulos-Lloyd, P. (nd) 'Queer futurism: Denizens of liminality', Queer Nature, Available from: https://www.queernature.org/queer-futurism-denizens-of-liminality
Steeves, P.F. (2015) 'Decolonizing Indigenous histories, Pleistocene archaeology sites of the Western hemisphere' [doctoral dissertation], Binghamton University.
Stewart-Harawira, M. (2005) *The New Imperial Order: Indigenous Responses to Globalization*, Zed Books.
Suchet-Pearson, S., Wright, S., Lloyd, K., Burarrwanga, L., and Country, B. (2013) 'Caring as Country: Towards an ontology of co-becoming in natural resource management', *Asia Pacific Viewpoint*, 54(2): 185–97.
Terry, G., Hayfield, N., Clarke, V., and Braun, V. (2017) 'Thematic analysis', in C. Willig and W.S. Rogers (eds) *The SAGE Handbook of Qualitative Research in Psychology*, SAGE Publications, pp 17–36.

Tsing, A.L. (2017), 'A threat to Holocene resurgence is a threat to livability', in M. Brightman and J. Lewis (eds) *The Anthropology of Sustainability: Beyond Development and Progress*, Palgrave Macmillan, pp 51–65.

Walton, K.L. (1993) *Mimesis as Make-Believe: On the Foundations of the Representational Arts*, Harvard University Press.

Whyte, K.P. (2018) 'Indigenous science (fiction) for the Anthropocene: Ancestral dystopias and fantasies of climate change crises', *Environment and Planning E: Nature and Space*, 1(1–2): 224–42.

Wildcat, D. (2009) *Red Alert! Saving the Planet with Indigenous Knowledge*, Fulcrum.

Wiltsher, N., and Meskin, A. (2016) 'Art and imagination', in A. Kind (ed) *The Routledge Handbook of Philosophy of Imagination*, Routledge, pp 179–91.

Yaszek, L. (2006) 'Afrofuturism, science fiction, and the history of the future'. *Socialism and Democracy*, 20(3): 41–60.

Young, L. (2023) '"The end of the world as we know it": Imagining new possibilities for the Anthropocene through a study of Nigerian Africanfuturism' [master's thesis], University of Cape Town, Available from: http://hdl.handle.net/11427/39938

4

We Need to Honour Emotions: Encoding Pathos to Make Science Communication Work as Sustainability

Sabrina Vitting-Seerup

In the so-called 'science of science communication', the goal has long been to transmit knowledge most effectively. You might expect me to now critically analyse why that is a terrible goal, or why it has never been achieved by the many sustainability scientists, science communicators, and activists working on effectively enlightening the public about our impending doom – but I won't. Because in many ways they succeeded.

The people of the Western World (and perhaps of the entire globe) know that there is a climate crisis created by humans (Taddicken et al, 2018; Ritchie et al, 2024). As Hans Rosling explained in his last book before passing, sustainability science has been able to disseminate the fact of climate change quite effectively – to a point where it is one of the most widely known scientific facts (Rosling et al, 2019). Furthermore, the support for climate action is vastly underestimated: 69 per cent of the world's population is reportedly ready to donate 1 per cent of their income to mitigate climate change (Andre et al, 2024).

People know.
They want to contribute.
Then why are we stuck in a collapsing world?

This is usually my cue for delivering yet another hit to the 'oil lobby', or for writing a structural critique of capitalism and democracy and their interrelations in the past 100 years … I have done both more than once, but I want to point to another problem in this particular chapter.

While people do *know* about the sustainability crisis, they do not want to think about it. Not because they don't see it as important, but because it engenders negative emotions, as seen in Kurth and Pihkala's taxonomy (2022). Emotions like fear can cause people to distance themselves from the climate crisis (O'Neill and Nicholson-Cole, 2009), and if science communicators attempt to push past the fear and demand that people listen using a more aggressive communication style, they risk evoking psychological reactance (Yuan and Lu, 2020).

Why shouldn't people feel bad, you might ask, pointing to excellent works on, for example, dark pedagogy (Lysgaard et al, 2019). Specifically, this line of research encourages us to 'say it as is', even when hope is nowhere to be found. In response, I recall a co-creation workshop I helped facilitate in 2023 (described in Achiam et al, 2024). In this workshop, researchers, communicators, teachers, consultants, artists, and activists discussed the challenges of working with sustainability science communication. We discussed climate anxiety, climate depression, and in particular, how nobody wanted to scare the children. 'But this is an emergency', an activist/PhD student said, arguing how too much hope would allow people to live their unsustainable lives in peace, leaving the Earth shattered. 'We need to be scared and *should* be depressed', a very dry academic agreed. But then a teacher stood up, frustration seeping out of her pores: 'when has depression *ever* solved a problem?' Most academics looked down in silence – including myself.

I am not suggesting that sustainability science communication should be all about hope and positivity. Rather, I want to highlight just how deeply emotions shape this field, and how we must engage with emotions if our science communication is to function *as* sustainability. To transform both mind and action, we need to feel, as Hochschild argues in her seminal work on emotions and labour. She states that emotions are 'related not only to an orientation toward action, but also to an orientation toward cognition' (Hochschild, 2003, p 229). Even though this connection from emotions to cognition ought to grant them a place in the field of science communication, the so-called 'emotional turn' in science communication has largely focused on audience responses (Bilandzic et al, 2020; Taddicken and Reif, 2020). There is, however, a growing recognition that emotions are not merely reactions to science but rather embedded within it (Head and Harada, 2017).

In this chapter, I want to honour the emotions inherent in sustainability science with the goal of fostering a science communication approach that prioritizes transformation over transmission. To do so, I will first explore why emotions seem so hidden within science, before looking at which emotions emerge from sustainability science itself. Then, I will consider how these emotions can affect the communication of sustainability science and how we might strategically engage with emotions to reach our aim of transformation. I do not seek to impose a hierarchy of emotions. Just as we would not privilege one scientific finding over another, I do not argue that hope is better than anger or fear better than joy. My aim is to encourage the diverse community of science communicators to value emotions as much as reason when working with science communication as sustainability, and my road to do so goes through recognizing emotions as an integral part of sustainability science itself.

Forbidden emotions

Sustainability scientists conduct research in a field that focuses on the collapse of ecosystems, on the transgression of nearly all planetary boundaries, on the heating climate's effect on life as we know it ... How is it possible to conduct research on such areas without feeling? And yet, scientists are expected to 'keep their heart a long way from the brain', as Head and Harada (2017) so poetically describe it.

For over a century, emotions have been largely unwelcome in science. In the early 1900s, researchers were required to detach themselves from the contexts they studied, reinforcing the notion that scientific enquiry requires distance (Simmel, 1964). This notion was codified in an ideal of objectivity in science that called for scientists to keep themselves and their emotions far away from what they were studying. Later came the CUDOS norms[1] which have effectively been shaping universities world-wide (Macfarlane and Cheng, 2008). The CUDOS norms are based on Merton's sociological study of scientific institutions, his original work defining four core principles of scientific practice. One of Merton's original principles was 'disinterestedness' (Merton, 1973). Despite critique of using Merton's work to describe contemporary academia (Scheufele and Krause, 2019), the ideal of the disinterested scientist seems to remain deeply embedded in academic culture. But how is that even possible? The researchers I know, including myself, have a burning passion for what we do. We are deeply invested in our subject – whether it is hedgehog mating patterns, protein isoforms, or how emotions might help us move sustainability communication from transmission to transformation.

Even outside my anecdotal evidence, scientists do seem to experience 'interest' and 'attachment' in connection to their research. Climate scientists, for instance, have been shown to be 'extremely worried about climate change and the future of the planet', yet they often practise 'self-censorship when engaging in the climate debate' (van Eck et al, 2025). This suppression is not accidental but can be seen as a direct product of their scientific training and the norms and ideals embedded within it, as scientists are actively taught to emphasize 'scientific rationality' (van Eck et al, 2025) and to downplay their emotions to maintain credibility (Head and Harada, 2017). Emotions are simply positioned as the opposite of reason and are therefore banned from science (Barbalet, 1998, p 29). This expectation of emotional suppression can also be traced back several decades, for example to Max Weber's reflections on science as a vocation. Weber emphasized the role of rationality in scientific enquiry (1970, p 141) and argued that true rationality is found in voluntary, free-willed action – actions free from emotions of any kind (1978, pp 23–4). Simmel, too, linked rationality to the control of emotions, though with a different emphasis. He suggested that rationality does not simply exclude emotions but emerges through their containment (in Barbalet, 1998, p 55). This emphasis on suppression seems to still influence science today, meaning that scientists rarely express their emotions in their scientific work or when communicating it. Is it a wonder that the illusion persists that science and scientists are completely void of emotions?

This expectation of detachment is particularly evident in academic writing, where emotional expression is not just discouraged but actively rejected. Scientists are trained to communicate objectively and neutrally (Douglas, 2009), often adopting what Grünfeld describes as a mode of 'disengaged engagement' (2019, p 49). Over time, this suppression makes 'scientists begin to resemble machines,' as Grünfeld notes through de Bonald (Grünfeld, 2019, p 44). Since we live in the era of 'publish or perish' academia (Yeo et al, 2021), meaning that we as scientists must publish in high-ranking academic journals to (continue to) be employed by a university, we are forced to fit our science into these emotionless formats.

Emotions in science have been treated as something to be controlled, suppressed, distanced, or hidden, but as sustainability science forces us to confront urgent, deep crises, it is becoming harder and harder to uphold this detachment … or perhaps it is just meaningless?

'We have been trying to warn you'

On a spring day in April 2022, Dr Peter Kalmus became front page news as he and a team of other scientists chained themselves to the doors of a JPMorgan Chase building in protest at fossil fuel investments: 'I'm here because scientists are not being listened to. I'm willing to take a risk for this gorgeous planet', Dr Kalmus said while chained to the door. He then started to cry as he added: 'For my sons' (Freedman, 2022).

'We have been trying to warn you', he explained. 'We're going to lose everything. And we're not joking, we're not lying, we're not exaggerating' (Freedman, 2022). Tears were running down his face but not muffling his words in any way. The barrier between reason and emotion seemed long gone.

Dr Kalmus is not an outlier. On that same day, scientists and fellow supporters protested at the Ministry of Environment in Quito, Ecuador – and in my hometown of Copenhagen. On social media, I saw colleagues obstructing traffic near the Danish Climate Ministry, refusing to move. I sat on the couch and watched research colleagues be carried away by police officers as I hugged my daughter a little tighter, both relieved and embarrassed that I didn't join the protest.

When sustainability scientists *do* express emotions, it is often in situations that are publicly categorized as activism – rarely are they described as public debate or science communication. Since activism is viewed as something to be detached from 'the actual science' (Olsen, 2023), the illusion persists: activism might be emotional, but science is rational and detached – even when tears stream down the face of a climate scientist talking about climate science. It brought to my mind a quote from Hochschild: 'A man who feels fear of a rattlesnake moving toward him may run to safety. He may act rationally. … The only reason I pose obvious examples is that when people talk about 'acting emotionally', it is often not these examples they cite' (Hochschild, 2003, p 214). There is a rise in the number of sustainability scientists becoming activists in attempts to stop the catastrophe they see moving closing and closer (Brysse et al, 2013). They do so despite knowing that their activism can and will be used as an argument against their scientific ethos (Olsen, 2023) – indeed, no one talks about the rationality of running away from a rattlesnake.

When sustainability scientists include emotions in their communication, it actually increases climate engagement (van Eck et al, 2025). Van Eck et al describe this shift as a tendency for scientists to transition 'from primarily factual and neutral communication to expressing their genuine feelings about climate change' (2025, p 2). But for those sustainability scientists who do not take to the streets, where do these emotions go?

Some sustainability scientists turn to alternative outlets – places where they can state freely what they cannot write in scientific papers. The project 'This Is How Scientists Feel' (www.isthishowyoufeel.com) provides one such space. Here, scientists write deeply personal letters to their loved ones about what it feels like to witness environmental destruction through the lens of data and research. The site reminds visitors before they open a letter: '[T]hey're not robots. These scientists are mothers, fathers, grandparents, daughters. They are real people. And they're concerned.' Why would we need to remind people that scientists are not robots, I wonder?

No room for feelings

When turning to the literature on how sustainability science is and should be conducted, I struggle to find the emotions. Still, I find it hard to believe that any sane human could research the collapse of the world around them and feel nothing …

Spangenberg (2011) describes how science *for* sustainability (meaning research conducted within discipline-specific academic frameworks) was, at the time of writing, the dominant form of sustainability science. Largely housed within universities, this type of research is often difficult to translate into real-world applications. Spangenberg argued that we have to move towards a science *of* sustainability that embraces interdisciplinary and transdisciplinary approaches (Spangenberg, 2011). He called for a move from 'hierarchical logic' to 'relational logic' (Spangenberg, 2011) – but would that leave room for emotions to be acknowledged, I wonder?

Spangenberg's perspective recognizes sustainability science as a field that grapples with wicked problems, that is, complex and systemic challenges that sit at the intersection of science and society (Dillon, 2017). Wicked problems, crucially, cannot be solved by science and technology alone (Dillon et al, 2021). They require diverse knowledge systems and perspectives (Achiam et al, 2021), including the emotional and affective dimensions that shape human decision making. And yet, even when we acknowledge that emotions play a role in sustainability science, we seem hesitant to name them explicitly – to give them a voice in our scientific endeavours.

Are emotions so banned by the norms of science that even when scientists argue for their importance, we shy away from sharing them? Is recognizing scientists as part of the world they study – as entangled in the world's crises rather than objective observers – still too radical a notion? I got stuck thinking about what would happen if sustainability scientists allowed themselves to be fully present in their work – not just as researchers, but as whole human beings ... as citizens, as activists, as artists, as parents? What if they suddenly just stopped hiding the way climate change breaks their hearts? Would it break the science? Or strengthen it?

Of course, there are academic traditions where emotions are somewhat recognized as valuable data. Ethnology and autoethnography, for example, are built on the idea that subjectivity is not a flaw but a source of insight. Yet, these approaches remain marginalized within the dominant landscape of sustainability science, which is largely housed within the natural sciences (Schoolman et al, 2012). It also seems that despite widespread acknowledgement that the wicked problems of sustainability require interdisciplinary approaches, knowledge transfer between neighbouring subfields happens only in small steps (Porter and Rafols, 2009) and has previously been disregarded as 'a bridge too far to cross' for any field (Leydesdorff and Schank, 2008, p 1817).

If we were to (1) view emotions as already embedded in sustainability science, as I have argued they are, and (2) acknowledge that mainstream science does not view those feelings as a productive outcome, and finally (3) consider how easily scientists' emotional communication gets downplayed as unscientific activism – then science communication might be one of the few arenas where we can paint the whole picture. The question then becomes: how do we honour the emotions of sustainability and present them as part of sustainability science rather than as unscientific add-ons?

Encoding – decoding – encoding

While looking for a fitting model to help me identify and work with the emotions embedded in sustainability and its science, I was advised by my good colleague Grünfeld (a co-editor and author of this book) to take a step back and question how emotions occur.

Research has long shown that emotions are neither purely internal occurrences nor entirely shaped by external forces. Rather, they exist at the intersection of biological structures, socialization, and learning processes (Emde, 1984; Lewis and Saarni, 1985). In other words, emotions are not simply experienced from within or imposed from without; they emerge through interactions and are shaped both by individual experience and broader societal contexts. But what kind of communication model can account for that? How do we analyse emotions as not only subjective experiences but also elements that are encoded into and decoded from communication events, influenced by the topic, academic norms, scientists' experiences, the science communicators' aim, and the audience's expectations – just to name a few?

My suggestion would be to start with Stuart Hall's encoding–decoding model, which provides a way to analyse the emotions encoded in sustainability science and its communication (consciously or not) and how these emotions might be interpreted by audiences, by taking into account the broader context of a communication event. Understanding these complex dynamics is crucial if we are to move beyond simply acknowledging that emotions exist in sustainability science and move towards actually *working with* emotions in our science communication efforts.

In Hall's article 'Encoding and decoding in the television discourse' (1980), he sought to move beyond the rigid, linear transmission models that had long dominated his field – much like what we aim to do in this book. Hall pointed out that if meaning could easily be perfectly transmitted, 'communication would be a perfectly equivalent circuit, and every message would be an instance of "perfectly transparent communication"' (Hall, 1980, p 52). Of course, it is not – and Hall's model shows how not just the production of a communication event but also the reception of it forms meaning. Hall, in other words, is treating both creating and receiving communication as active and constructive processes. At the same time, Hall made it clear that 'relations and conditions' (Hall, 1980, p 48) shape all communication, giving context a decisive role in meaning making. This means that both the creation (encoding) and the reception (decoding) of a message are shaped by various factors, including the communicator's intentions, the audience's background, the broader societal and discursive contexts, and so on. This perspective complicates any straightforward notion of communication, particularly in fields like science communication, where clear messaging and measurable impact have been seen as essential. Still, I hope we dare face this more complicated route, as it will ultimately allow us to engage more honestly with the emotional dimensions of sustainability science, and in doing so, communicate not only facts but meaning, urgency, and care. I truly believe this is one route we can take if we want to move our science communication from transmitting already-known facts to encouraging much-needed transformations.

Hall's model also allows us to examine the multiple interpretations (and connected emotions) that emerge when encoding and decoding do not align perfectly (Hall, 1980, p 47). In Hall's view, communicators *can* shape messages to encourage specific interpretations (what he calls 'preferred readings'), but communication will always contain polysemic values – or multiple possible meanings (Hall, 1980, p 50). While most communication events have preferred readings that are encoded by the communicator(s), these readings are rarely the only messages conveyed. Rather than assuming a message will speak for itself, Hall urged us to analyse 'the transposition into and out of the "message form"', emphasizing that encoding and decoding are not random processes that can be ignored or controlled at will (Hall, 1980, p 46). They are fundamental to understanding how any message – scientific or otherwise – is received, interpreted, and emotionally engaged with.

Applying Hall's model to science communication as sustainability allows us to explore not just *which* emotions are present in sustainability discourse but also *how* they are encoded and decoded within different contexts. It offers a means to analyse how conditions and relations affect our science communication, and perhaps which emotions seep in from scientists, from audiences, and from ourselves as science communicators. Another hope I have for implementing the encoding–decoding model is that it frees us from focusing solely on whether something is a 'good' or 'bad' feeling in terms of fostering sustainability, allowing us instead to see emotions in science communication as more than a one-stringed instrument. By using Hall's model and his focus on polysemic values, it becomes clear that many different emotions will arise in a decoding and that the 'preferred' emotions to encode in your communication event depends on the message and contexts you are working with – the terroir of your science communication, we might say (Chapter 2).

Hall's model has many qualities, but it only allows us to analyse what is already there at a somewhat manageable complexity. If we wish to work with emotions as strategically as we work with rationality in science communication, we need a language and an approach for how to appeal to *both* reason and emotion, and how to balance the two.

Encoding appeals

As soon as the word 'appeal' entered my mind, I thought of Aristotle. He wrote about appeals at a point in time where modern science was not yet created, and the climate crises were not even imaginable. Even so, Aristotle's focus on striking a necessary balance between reason and emotion might be just what we need. Especially because his work on the one hand gives us a straightforward definition and on the other leaves wiggle room for theorists to define and practitioners to operationalize.

It was around 350 BCE when Aristotle named the three appeals – what he calls 'modes of persuasion':

> Of the modes of persuasion furnished by the spoken word there are three kinds. The first kind depends on the personal character of the speaker; the second on putting the audience into a certain frame of mind; the third on the proof, provided by the words of the speech itself. (Aristotle, 2001, 1.2.3)

Aristotle goes on to name the three modes of persuasion, defining ethos as the personality and/or trustworthiness of the speaker, pathos as an appeal to the audience's emotion, and logos as rationality-based reasoning. Interestingly, as Aristotle moves on to describe these modes of persuasion, he divides ethos into three distinct qualities: *phronesis* (signalling 'good sense', for example intelligence and wisdom), *arete* (signalling moral virtue), and *eunoia* (signalling goodwill [Aristotle, 2001, 2.1.5 and 1.8.6]). Even more attention is given to defining logos, which is presented as the superior form of argument in that it holds *enthymeme* (often translated to 'rational deduction') and *paradigm* (meaning 'proof through example') (Aristotle, 2001, 1.2.8). Pathos, in contrast, is described as simply 'putting the listener into a certain frame of mind' (*ton akroaten diatheinai poos* [Aristotle, 2001, 1.2.3]). As Braet describes in his re-examination of the ancient text, Aristotle viewed pathos (and ethos) as 'non-argumentative and inferior forms of persuasion' compared to logos (Braet, 1992, p 309). Yet, Aristotle found deep importance in 'putting the listener into a certain frame of mind', advocating that the speaker should be able to 'understand the emotions' and 'name them and describe them, to know their causes and the way in which they are excited' (*Rhetoric*, book I, chapter II). How many science communicators attend to these details, I wonder? How many have named 'activist anger' or 'sustainability euphoria'[2] and treated that naming with as much attention as the references on their final slide?

Today, pathos is broadly defined as 'an appeal to human emotion' (Crowley and Hawhee, 2004, p 205) and is firmly placed in the field of rhetorical studies alongside the other two Aristotelian appeals. The three are still mainly used to examine and design speech acts as Aristotle intended – but are also used beyond speech. Higgins and Walker define rhetorical studies as being 'concerned with how language and *other symbolic forms* influence the way an audience thinks, feels or acts' (2012, p 197; my italics). Literary theorist and philosopher Kenneth Burke famously wrote: 'Wherever there is persuasion there is rhetoric. And wherever there is "meaning" there is "persuasion"' (Burke, 1969, p 172). Burke thereby identifies a vast scope for rhetoric. Foss et al (2002) have pointed out that Burke's views on rhetoric are not always seen as being in accord with those of Aristotle; even so, parallels have been noted between Burke's terminology and the three Aristotelian appeals (Schwartz, 1966; Higgins and Walker, 2012).

If we jump to science communication and sustainability studies, researchers are already using Aristotle's appeals to analyse the dissemination of sustainability science in many different formats. As early as 1990, Waddell used the appeals to analyse sustainability policies. Higgins and Walker (2012) have used the three appeals to understand the persuasive tools implemented in sustainability reports. Molek-Kozakowska (2017) and Voci (2022) have looked at the ways sustainability science journalism uses the appeals, and Northcut (2006) studied how the appeals were expressed in images used in sustainability science journalism. Also, Simon (2020) and Taddicken and Wolff (2020) have shown that all three appeals can be found in public discourse discussing sustainability, for example on social media.

In these studies, the focus tends to be on the appeals' broad ability to persuade – which makes sense, as Aristotle described the three appeals as 'modes of persuasion'. An example is Manzo (2017), who uses all three appeals in her meta study of climate change films and documentaries as a tool to uncover the 'purposeful education' found in climate communication and its focus on 'moving people to take action' (Manzo, 2017). In this view, sustainability communication aims not simply to *inform* the audience but to *persuade* them – a clear sign that using pathos as a tool to analyse science communication comes with certain understandings. The question is whether persuasion is enough to reach the transformation we are working towards in science communication as sustainability.

Perhaps it is not persuasion in itself that might pose a problem but what we are we trying to persuade our audiences of? Or, in other words, if we stop aiming to persuade audiences that our facts and statistics are correct, and instead tried to persuade our audiences that different ways of knowing have value, that other ways of life can be celebrated, that this world holds many lifeforms that we need to protect and treasure … Maybe such persuasions would lead to transformations in ways that persuasive empirical evidence cannot? For that, however, we need another language than that we have been using in sustainability science communication up until now – one that isn't found in academic journals and on the TED stage. A language that could balance appeals, work with multiple decodings and contexts – a language that would not just persuade us of findings but let us imagine new ways of being …

I think it is time we turn to the arts.

Encoding pathos into arts-based approaches

I don't just turn to the arts because I want to. Multiple openings throughout this chapter have guided me towards this point:

> If our goal is to leave the 'disengaged engagement' of science behind …
> If our goal is not to solely create effective alignment between the encoded and decoded …
> If our goal is to appeal to emotions *and* logic – to persuade our audiences that another way of life is possible …
> then arts-based approaches might be just what we need.

I have actually tried to avoid arts-based approaches in my research. I deemed them silly. Or more precisely: I was afraid that others would deem my research silly if I dared work with arts-based methods, and so I left them behind and turned to critical discourse and thematic analyses. My work is already easily disregarded as naïve or activist, seeing as my research has been focusing on questions of inclusion and sustainability. 'How topical,' an older associate professor once told me with a smirk across his face. 'Are you aiming for a career in ESG?'[3] His question of whether I intended to work in industry instead of academia was *not* meant as a compliment, I assure you.

In retrospect, it feels like I was led back to the arts, even though I tried to take alternative routes. My bachelor's degree is in comparative literature, and my master's thesis was on comics' ability to create empathy among readers, so perhaps I shouldn't be too surprised? And yet, I didn't imagine this would be where my work led me. As a researcher valuing honesty and transparency, I wanted you to know that as you read the end of this chapter.

The arts have an ability to create conditions for both the concrete and the resonant (Rosa, 2019, p 284). Arts-based formats, such as visual arts, storytelling, music, and performance activate both sensory and emotional dimensions of knowledge, fostering an engagement that is not just intellectual but embodied and felt. The language of the arts leaves room for emotions and footnotes, for people and critiques, for visuals, sound, and imaginaries beyond our first comprehensions. Where traditional science communication often relies on transmitting methods, facts and data, the arts translate insights that can be felt and sensed. They provide affective entry points – moments where audiences form personal and emotional connections to sustainability challenges (Heras et al, 2021). The arts can foster empathic engagement, making abstract or distant issues feel more immediate, relevant, and actionable (Bennett, 2005).

Crucially, learning from the language of the arts can help us expand the emotional range of our science communication. As of right now, we seem mostly affected by negative emotions when discussing sustainability (Brosch, 2021; Kurth and Pihkala, 2022), but anxiety and anger are not the only emotions available within this field. Sustainability messages can evoke positive emotions, like optimism, pride, and gratitude, depending on the framing and the stories that are told (Schneider et al, 2021). Humour and satire might cut through eco-fatigue, and beauty and awe can rekindle a sense of care for the natural world.

Beyond shaping emotions, the arts can create transformative spaces – places where audiences are invited to participate in the imagining of new worlds. In the words of art historian Meskimmon: 'Art does more than reflect the world – it produces it' (Meskimmon, 2013, p 41). Through arts-based experiences, people can begin to envision new futures, question existing structures, and imagine alternative ways of living. Wouldn't it be wonderful if they did so while standing on the bedrock of science?

Importantly, the arts thrive on ambiguity and multiple interpretations, which makes them well suited for engaging with sustainability's wicked problems. Where the science of science communication often prioritizes clarity and definitive conclusions (Canon et al, 2022), arts-based approaches invite dialogue, questioning, and a capacity to hold contradictions (Heras et al, 2021). The open-endedness of arts-based approaches thereby aligns closely with Hall's encoding–decoding model which reminds us that communication is never neutral and always co-constructed between encoder and decoders. The language of the arts can make this explicit by embracing multiple perspectives in their formats. They challenge the notion that sustainability science must always be conveyed through detached, objective language, instead offering ways to acknowledge emotions, positionality, and lived experience. If sustainability science communication is to move from transmission to transformation, then borrowing elements from the language of the arts might not just be useful – it might be essential.

At this point, I probably have to clarify that by borrowing from the arts I do not intend to turn science communication into art (or even worse: bad art). The goal remains to create science communication that functions as sustainability by balancing pathos with logos – hopefully without losing our ethos along the way. Learning from the arts does not require scientists or science communicators to become artists, nor does it mean that all science communication must take on an artistic form. Rather, it means that embracing artistic strategies – such as storytelling, metaphor, or immersive experience – can help us develop new tools to strengthen our science communication while maintaining scientific integrity.

Arts-based approaches can even provide us with more than a language, as it points towards a new epistemological scope. Moser (2019) describes sustainability science communication as 'transformative work and itself transformational', arguing that it must be moving, poetic, and future oriented. She frames this kind of communication as 'a form of creating the future', positioning sustainability science and the arts in an inherently intertwined relationship. In this sense, the arts do not merely support science; they expand its reach, integrating complexity, nuance, and new ways of knowing. Rather than positioning emotions as secondary to rational analysis, arts-based approaches recognize that emotions are fundamental to how people understand, connect with, and respond to sustainability issues. Heinrichs (2018) also highlights that many transdisciplinary models fail to incorporate diverse knowledge claims, but in the arts, affect and lived experience can unfold next to more traditional expressions of science. Heinrichs and Kagan (2019) further argue that sustainability science must expand its epistemological approaches to account for the multisensory and aesthetic dimensions of human life, if the sciences are to truly study sustainability (or lack thereof) in all its forms.

To be clear, arts-based science communication should not abandon reason or rigor. If it slips into mere sentimentality, this kind of science communication risks losing its transformative potential. I simply argue that the balance between emotion and reason probably can't be found in fixed models but needs to be discovered through experimentation performed by science communicators daring to move from explanation to evocation. Theorists can support this by offering tools to identify the emotions already encoded in sustainability science, and by helping shape a language for new formats. This work will require courage and all kinds of collaborations. Perhaps that means going back to old traditions in new ways (Chapter 7), or that we sit down and start making things together (Chapter 6). Most likely, it will imply many things that will only begin to reveal themselves as we start to practise what we preach.

Implications for science communication as sustainability

Ultimately, honouring emotions in science communication as sustainability means recognizing the rightful place of emotion alongside knowledge in the urgent task of fostering transformation. After all, as Hochschild reminds us, emotions are 'our experience of the body ready for an imaginary action' (2003, p 230). If science communication as sustainability is to be more than a chronicle of collapse, we must be willing to communicate not just what we *know* but also what we *feel* and invite others to feel it with us.

The shift I advocate for in this chapter requires an awareness of what is being encoded and decoded, a commitment to balancing emotional engagement with scientific findings, and a willingness to borrow language from other ways of communicating – like that of the arts. To integrate emotions more intentionally into science communication as sustainability, we can ask ourselves:

- *What emotions are already present?* Before crafting a message, reflect on the emotions that are already embedded in your topic, in the room, in your own body – perhaps with the help of Hall's encoding–decoding model? Build your science communication from there rather than trying to hide it.

- *How can we be transparent about the role of emotions in sustainability science?* Are we acknowledging that emotions influence both researchers and audiences, shaping how sustainability knowledge is produced, communicated, and understood? Or do we pretend the emotions are not there to live up to a set of outdated norms?

- *What do we* want *the audience to feel?* If transformation is the goal, which emotions might motivate rather than paralyze? How do we balance urgency with agency?

- *How to appeal to the preferred emotions?* Are we relying on traditional formats that privilege detached logos appeals, or can we perhaps incorporate pathos and imagination with the help of arts-based approaches such as storytelling or music?

- *Are we creating space for multiple interpretations?* Rather than assuming our preferred readings will reach the audience as a transmission, how can we encourage audiences to participate, question, and co-create meaning? And how do we, as science communicators, live with the somewhat uncontrollable outcome?

Notes

1 CUDOS refers to Robert K. Merton's classic normative principles for science and is an acronym for the words: Communalism, Universalism, Disinterestedness, and Organized Scepticism (Merton, 1973).

2 'Sustainability euphoria' – meaning the excitement, joy, and sense of possibility that come from engaging with sustainable futures. And yes, that is a word I just made up because it needs to exist.

3 ESG stands for Environmental, Social, and Governance. The abbreviation is widely used in Scandinavian industry, where the three foci work as a guide for evaluating how a company behaves in these three key areas.

References

Achiam, M., Glackin, M., and Dillon, J. (2021) 'Wicked problems and out-of-school science education: Implications for practice and research', in M. Achiam, J. Dillon, and M. Glackin (eds), *Addressing Wicked Problems through Science Education: The Role of Out-of-School Experiences*, Springer, pp 229–37.

Achiam, M., Vitting-Seerup, S., Whiteley, L., and Dam, S.L. (2024) 'Creating resonance with arts-based approaches to sustainability science communication', *Journal of Science Communication*, 23(6): N01.

Andre, P., Boneva, T., Chopra, F., and Falk, A. (2024) 'Globally representative evidence on the actual and perceived support for climate action', *Nature Climate Change*, 14: 253–9.

Aristotle (2001) *Rhetoric*, in P. Bizzell and B. Herzberg (eds) *The Rhetorical Tradition: Readings from Classical Times to the Present* (2nd edn), Bedford/St. Martin's, pp 179–240.

Barbalet, J.M. (1998) *Emotion, Social Theory, and Social Structure: A Macrosociological Approach*, Cambridge University Press.

Bennet, J. (2005). *Empathic Vision: Affect, Trauma, and Contemporary Art*, Stanford University Press.

Bilandzic, H., Kinnebrock, S., and Klingler, M. (2020) 'The emotional effects of science narratives: A theoretical framework', *Media and Communication,* 8(1): 151–63.

Braet, A.C. (1992) 'Ethos, pathos and logos in Aristotle's Rhetoric: A re-examination', *Argumentation*, 6(3): 307–20.

Brosch, T. (2021) 'Affect and emotions as drivers of climate change perception and action: A review', *Current Opinion in Behavioral Sciences*, 42: 15–21.

Brysse, K., Oreskes, N., O'Reilly, J., and Oppenheimer, M. (2013) 'Climate change prediction: Erring on the side of least drama?', *Global Environmental Change,* 23: 327–37.

Burke, K. (1969) *A Rhetoric of Motives*, University of California Press.

Canon, C.R., Boyle, D.P., and McAfee, S.A. (2022) 'Visualizing the structure and development of climate change communication research', *Journal of Science Communication*, 21(7): A03.

Crowley, S., and Hawhee, D. (2004) *Ancient Rhetorics for Contemporary Students* (3rd edn), Pearson/Longman.

Dillon, J. (2017) 'Wicked problems and the need for civic science', *Spokes*, 29: 1–9.

Dillon, J., Achiam, M., and Glackin, M. (2021) 'The role of out-of-school science education in addressing wicked problems: An introduction', in M. Achiam, J. Dillon, and M. Glackin (eds) *Addressing Wicked Problems through Science Education: The Role of Out-of-School Experiences*, Springer, pp 1–8.

Douglas, H.E. (2009) *Science, Policy, and the Value-Free Ideal*, University of Pittsburgh Press.

Emde, R.N. (1984) 'Levels of meaning in infant emotions: A biosocial view', in K.R. Scherer and P. Ekman (eds) *Approaches to Emotion*, Lawrence Erlbaum Associates, pp 77–107.

Foss, K.A., Foss, S.K., and Trapp, R. (eds) (2002) *Readings in Contemporary Rhetoric*, Waveland Press.

Freedman, A. (2022) '"We've been trying to warn you": NASA climate scientist breaks down in tears at protest', *The Independent* [online], 13 April, Available from: www.independent.co.uk/climate-change/news/nasa-scientist-protest-extinction-rebellion-b2057563.html

Grünfeld, M. (2019) *Writing and Thinking in Contemporary Academia: The Poetics of Clarity*, Routledge.

Hall, S. (1980[1973]) 'Encoding/decoding', in S. Hall, D. Hobson, A. Lowe, and P. Willis (eds) *Culture, Media, Language*, Hutchinson, pp 128–38.

Head, L., and Harada, T. (2017) 'Keeping the heart a long way from the brain: The emotional labour of climate scientists', *Emotion, Space and Society*, 24: 34–41.

Heinrichs, H. (2018) 'Sustainability science with Ozzy Osbourne, Julia Roberts and Ai Weiwei: The potential of arts-based research for sustainable development', *GAIA: Ecological Perspectives for Science and Society*, 27(1): 132–7.

Heinrichs, H., and Kagan, S. (2019) 'Artful and sensory sustainability science: Exploring novel methodological perspectives', *Revista de gestão ambiental e sustentabilidade*, 8(3): 431–42.

Heras, M., Galafassi, D., Oteros-Rozas, E., Ravera, F., Berraquero-Díaz, L., and Ruiz-Mallén, I. (2021) 'Realising potentials for arts-based sustainability science', *Sustainability Science*, 16(6): 1875–89.

Higgins, C., and Walker, R. (2012) 'Ethos, logos, pathos: Strategies of persuasion in social/environmental reports', *Accounting Forum*, 36(3): 194–208.

Hochschild, A.R. (2003) *The Managed Heart: Commercialization of Human Feeling* (20th anniversary edn), University of California Press.

Kurth, C., and Pihkala, P. (2022) 'Eco-anxiety: What it is and why it matters', *Frontiers in Psychology*, 13.

Lewis, M., and Saarni, C. (eds) (1985) *The Socialization of Emotions*, Plenum Press.

Leydesdorff, L., and Schank, T. (2008) 'Dynamic animations of journal maps: Indicators of structural changes and interdisciplinary developments', *Journal of the American Society for Information Science and Technology*, 59(11): 1810–18.

Lysgaard, J.A., Bengtsson, S., and Laugesen, M. (2019) *Dark Pedagogy: Education, Horror and the Anthropocene*, Palgrave Macmillan.

Macfarlane, B., and Cheng, M. (2008) 'Communism, universalism and disinterestedness: Re-examining contemporary support among academics for Merton's scientific norms', *Journal of Academic Ethics*, 6(6): 67–78.

Manzo, K. (2017) 'The usefulness of climate change films', *Geoforum*, 84: 88–94.

Merton, R.K. (1973) 'The normative structure of science', in R.K. Merton (ed) *The Sociology of Science: Theoretical and Empirical Investigations*, University of Chicago Press, pp 267–78.

Meskimmon, M. (2013) 'The promise of cosmopolitanism: Art, ethics and imagination in a global world – a positional parlay', in A. Macedo, C. de Sousa, and V. Moura (eds) *XIV Colóquio de Outono: Humanidades; Novos paradigmas do conhecimento e da investigação*, Edições Húmus, pp 33–46.

Molek-Kozakowska, K. (2017) 'Communicating environmental science beyond academia: Stylistic patterns of newsworthiness in popular science journalism', *Discourse & Communication*, 11(1): 69–88.

Moser, S.C. (2019) 'Not for the faint of heart: Tasks of climate change communication in the context of societal transformation', in G. Feola, H. Geoghegan, and A. Arnall (eds) *Climate and Culture: Multidisciplinary Perspectives on a Warming World*, Cambridge University Press, pp 141–67.

Northcut, K.M. (2006) 'Images as facilitators of public participation in science', *Journal of Visual Literacy*, 26(1): 1–14.

O'Neill, S., and Nicholson-Cole, S. (2009) '"Fear won't do it": Promoting positive engagement with climate change through visual and iconic representations', *Science Communication*, 30(3): 355–79.

Olsen, F. A. (2023) 'Forskeroprør ved Klimaministeriet: Urolig videnskabelig etos ved Scientist Rebellions teach-in-demonstration', *Rhetorica Scandinavica*, 27(86): 160–79.

Porter, A.L., and Rafols, I. (2009) 'Is science becoming more interdisciplinary? Measuring and mapping six research fields over time', *Scientometrics*, 81(3): 719–45.

Ritchie, H., Roser, M., and Sambroska, V. (2024) 'Climate change', Our World in Data [online], Available from: https://ourworldindata.org/climate-change

Rosa, H. (2019) *Resonance: A Sociology of Our Relationship to the World* (J.C. Wagner, trans), Polity.

Rosling, H., Rosling, O., and Rönnlund, A.R. (2019) *Factfulness: Ten Reasons We're Wrong about the World – and Why Things Are Better than You Think*, Sceptre.

Scheufele, D.A., and Krause, N.M. (2019) 'Science audiences, misinformation, and fake news', *Proceedings of the National Academy of Sciences*, 116(16): 7662–9.

Schneider, C.R., Zaval, L., and Markowitz, E.M. (2021) 'Positive emotions and climate change', *Current Opinion in Behavioral Sciences*, 42: 114–20.

Schoolman, E.D., Guest, J.S., Bush, K.F., and Bell, A.R. (2012) 'How interdisciplinary is sustainability research? Analyzing the structure of an emerging scientific field', *Sustainability Science*, 7(1): 67–80.

Schwartz, J. (1966) 'Kenneth Burke, Aristotle, and the future of rhetoric', *College Composition and Communication*, 17(5): 210–16.

Simmel, G. (1964) 'Types of social relationships (Original work published 1906)', in K.H. Wolff (ed) *The Sociology of Georg Simmel*, Free Press, pp 317–29.

Simon, N. (2020) 'Investigating ethos and pathos in scientific truth claims in public discourse', *Media and Communication*, 8(1): 129–40.

Spangenberg, J.H. (2011) 'Sustainability science: A review, an analysis and some empirical lessons', *Environmental Conservation*, 38(3): 275–87.

Taddicken, M., Reif, A., and Hoppe, I. (2018) 'What do people know about climate change – and how confident are they? On measurements and analyses of science related knowledge', *Journal of Science Communication*, 17(3): A01.

Taddicken, M., and Reif, A. (2020) 'Between evidence and emotions: Emotional appeals in science communication', *Media and Communication*, 8(1): 101–6.

Taddicken, M., and Wolff, L. (2020) '"Fake news" in science communication: Emotions and strategies of coping with dissonance online', *Media and Communication*, 8(1): 206–17.

van Eck, C.W., and van der Meer, T.G.L.A. (2025) 'Narratives of hope and concern? Examining the impact of climate scientists' communication on credibility and engagement', *Public Understanding of Science*, 34(6).

Voci, D. (2022) 'Logos, ethos, pathos, sustainabilitos? About the role of media companies in reaching sustainable development', *Sustainability*, 14(5): 2591.

Waddell, C. (1990) 'The role of pathos in the decision-making process: A study in the rhetoric of science policy', *Quarterly Journal of Speech*, 76(4): 381–400.

Weber, M. (1970[1919]) 'Politics as a vocation' (H.H. Gerth and C. Wright Mills, trans), in H.H. Gerth and C. Wright Mills (eds) *From Max Weber: Essays in Sociology*, Routledge, pp 77–128.

Weber, M. (1978[1921]) *Economy and Society: An Outline of Interpretive Sociology* (G. Roth and C. Wittich, eds), University of California Press.

Yeo, M.A., Renandya, W.A., and Tangkiengsirisin, S. (2021) 'Re-envisioning academic publication: From "publish or perish" to "publish and flourish"', *RELC Journal*, 53(1): 266–75.

Yuan, S., and Lu, H. (2020) '"It's global warming, stupid": Aggressive communication styles and political ideology in science blog debates about climate change', *Journalism & Mass Communication Quarterly*, 97(4): 1003–25.

5

We Need to Welcome Other Perspectives: Inviting Interference

Martin Grünfeld with Maria Brænder, Sofie Louise Dam, Adam Dickinson, and Alison Pouliot

In place of an introduction to this chapter, an invitation. An invitation to welcome other perspectives. Welcoming is an act of hospitality. It is an opening towards the other – an opening that might change you.

This chapter is placed in the middle of the book to create a passage towards more artistic and practice-based approaches in the second half of the book (curating in Chapter 6, historical imagination in Chapter 7, and listening in Chapter 8). The chapter is, however, also in itself a passage towards a welcoming of diverse perspectives even if they do not seem to fit neatly within established regimes of science communication. Indeed, this chapter is an exploration of the ways we engage art and science to consider what arts-based approaches to science communication might also sound like, look like, and feel like. As an opening into other perspectives, the chapter is explorative in its format ignoring what a chapter is supposed to do by including multiple voices and forms of expression: a poem, photographs, a sound excerpt from a performance, and a cartoon.

In this chapter, I have invited a performance and sound artist, a poet, a photographer, and a cartoonist to respond to the idea of science communication as sustainability. The people involved in making this chapter – the Danish performance and sound artist Maria Brænder, the Canadian poet Adam Dickinson, the Australian photographer Alison Pouliot, and the Danish cartoonist Sofie Louise Dam – have all graciously accepted my invitation and shared their perspectives. The chapter is based on individual conversations taking place between December 2024 and January 2025 with Adam, Maria, and Alison attended by Sofie and me. Each conversation was organized around a particular piece from their work acting as a focal point for discussing their practice and what we can learn from them within science communication.

When I invited Adam, Alison, and Maria to talk with me about science communication as sustainability, I already knew their work well. I perceive them as my like-minded kin in an increasingly complex and discomforting world. I deliberately invited them to contribute to this chapter because I knew that they were coming to science communication from different angles that would enrich our understanding of what it would mean for science communication as sustainability to performatively embody what it preaches (or more precisely how it performs rather than preaches). Yet it is worth noting that each conversation evolved in different and unexpected ways. From my initial invitation, their responses also involved a counter invitation to delve into their multifarious perspectives and learn from their practices even if it means unlearning the usual ways of understanding science communication.

In a moment, I will let Adam, Maria, and Alison and their works populate the pages. Here, however, I would like to emphasize that as part of the invitation to welcome other perspectives, they have individually selected which works to include in the chapter and written their own descriptions of their approaches in their own words. Hospitality is also always a matter of voicing – of who speaks and who is silenced. A crucial point in welcoming other perspectives is precisely to allow for multiple voices and approaches to unfold, instead of colonizing the text with one Master voice. Rather than a Master voice, I see my role as a curator who makes space and cares about their approaches. Following their works and thoughts, I respond with an interlude based on our conversations in relation to broader concerns in this book. As a last step in the chapter to cultivate connections between art and science, I have invited the cartoonist Sofie to make a comic based on our conversations with Adam, Maria, and Alison and her reception of the idea of science communication as sustainability.

Finally, I would like to welcome you – the reader – and your perspective. I encourage you to approach this chapter creatively and read it in whatever way you like.

Sincerely,
Martin Grünfeld

Adam Dickinson: poet and professor in English language and literature

Located at the intersection of poetry and science, my work explores how we 'write' the environment through pollution and industrial practices, and how the environment, in turn, 'writes' us through altered endocrine function and gut microbial health. By combining poetry with laboratory-based science, I investigate the porous boundary between inside and outside, tracing the metabolic exchanges that link bodies to global systems of energy and capital. Poetry, with its attention to the limits of language and signification, becomes a way to render these otherwise inscrutable inscriptions legible.

This literary approach offers an unconventional form of science communication by translating abstract or invisible biological processes into more intimate, cultural narratives, even as they become at times speculative and surreal. It potentially helps audiences perceive the embodied consequences of environmental change and encourages deeper engagement with issues like pollution and public health not just as data points, but as lived, affective realities. Additionally, however, I see my work not simply as a mode of communicating scientific information but as a conceptual, experimental practice that itself seeks to push the boundaries of knowledge. I am not aiming to create poetic ornaments of existing scientific ideas; rather, my goal is to merge literary and scientific experimentation to generate expanded forms of writing and thinking that might offer alternative ways of approaching scientific practices and problems alongside the broader social and cultural contexts in which they are embedded.

The accompanying poem – *Counter Film* – emerges from microbiology experiments that are part of a new book I am working on. *Terraforms* is a hybrid poetry book and bio-art project that unfolds from a double aftermath: the death of my father and the transformed biological landscapes of the Anthropocene. These personal and planetary ruptures collide in the interior terrain of the body, where grief and environmental disruption share strange chemical ground. The microbiome, that dense, fermenting rainforest of microbial life within the gut, becomes this uncanny meeting place. The microbes within us are ancient, symbiotic intelligences that have long co-authored human life, shaping thought, emotion, metabolism, and memory. They are also deeply responsive to environmental change, dietary shifts, antibiotic exposure, synthetic chemicals, grief, and stress. They are alien co-inhabitants that help regulate mood, immunity, even behaviour, yet their complex languages remain largely inscrutable.

Through a series of experimental encounters staged within a bioreactor seeded with a sample of my own gut microbiome, I explore how grief, pollution, and industrialized life disturb the body's internal ecosystem. In each experiment, I introduced substances such as antibiotics, microplastics, artificial food additives, soil from my childhood home, even my father's cremated ashes, to see how my microbial community responded to these 'new worlds'. These interventions terraformed the gut into a site of speculative ecology and intimate estrangement. The resulting poems in *Terraforms* are themselves terraforms, formal experiments that take microbial function and process as constraint and method as they respond to the perturbations in the microbial signals. *Terraforms* explores kinship with our inner aliens, engaging them not as threats but as guides through grief and ecological upheaval. What if the gut, our ancient intuitive organ, is already speaking in the language of the new worlds we must come to know?

In one experiment, I simulated an 'extinction event' by applying a cocktail of antibiotics to a steady-state sample of my gut microbiome. As expected, most strains of my resident gut bacteria declined in abundance. However, *Enterococcus hirae* appeared to thrive, likely by forming a biofilm, which is a slimy, protective matrix that bacteria use to survive stress. Biofilms resist antibiotics and adapt through collective defence. The poem, *Counter Film*, responds to biofilm as a sticky, subversive mode of resistance against imposed ideals of purity and control.

COUNTER FILM
We actively cultured
a cold

We gave it to each other
sharing

spoons, forks, and toothbrushes
holding

the gust of someone's sneeze
for as long

as we could like the shotgun
from a blunt

We brought remote controls
to our faces

doorknobs to our lips
shook hands

and hugged with infectious
abandon

Within us histamines
thickened

passages filled with expressive
currents

cisterns primed for our mission
We washed

in swamps and sloughs
and argued

on long walks with wet feet
exchanging

foundational tenets
of collaborative sweat

When they came for us
our noses were running

all at once
throats sore

with inflammatory pleats
a surface area

expanding, stretching
like our commitment

to the sovereignty of scum
and moss mouths

resisting in mucus mess
the gleaming counter

and polished mirror
of untangled light

from the new
drug

and its despotic threat
to never

make us sick
again

Alison Pouliot: ecologist, photographer and author

My work merges the science and the aesthetics of fungi. These remarkable organisms represent a diverse kingdom, and among its more curious members are the lichens. First, because they're old. Very old. They transformed primeval landscapes of rock to soil, and along with collaborators, enabled life on Earth. Lichens have persisted, almost unchanged throughout the ages, defying extremes and conditions intolerable to most other life, and hence the moniker, 'extremophiles' (see Figure 5.1). They're also stunningly beautiful. They create complex tapestries with other organisms with whom they form intimate affinities as cryptogamic crusts, stabilizing soils, creating microcosms, harbouring life (see Figure 5.2).

During many years of working in conservation, my notion of biodiversity has evolved. More than simply the scientific concept of the variety of life, or a product of evolution through deep time, 'biodiversity' is also a poignant aesthetic idea. Over the years of creating installations of fungi and guiding people through forests in search of them, I've observed how it's the aesthetics of fungi that people commonly respond to. Some might then become interested in their ecology or other aspects of their existence, but it's their exquisite beauty and bizarreness that ignite curiosity.

Melding the science and aesthetics of fungi offers an alluring entry point to their kingdom. The staggering diversity of fungus forms tempts us to explore

Figure 5.1: Opportunistic mosses and lichens form a tapestry on and among paving stones

Source: Alison Pouliot

Figure 5.2: A canal wall provides a canvas for mosses, lichens, algae, and others

Source: Alison Pouliot

the notion of biodiversity as key to complexity, robustness, and resilience. And not just biodiversity, but diversity in human societies, lifestyles, diets, music, art, gardens, everything. Concepts in ecology have grown in recent decades and have been influenced by other disciplines. Lichens prompt important questions, challenging us to go beyond 'sustainability' and think about resilience.

Fungi are unruly renegades. They don't play by the rules and operate at the edges. These interfaces are places of dynamism and innovation. Lichens occupy extreme and often specialist niches. They're pioneers, navigating the edges of biological limits, extracting moisture from a passing waft of mist or ingeniously producing sunscreens to resist solar radiation. By being collaborative, the lichen alliance broadens the range of conditions that individual players could withstand on their own.

We're just beginning to explore the creative potential of fungi. And I don't just mean, for example, using lichens to dye textiles; that's been happening for hundreds of years. But rather, the imaginative, creative possibilities for thinking about a different world, wobbling the assumptions of how we understand nature and life, and rethinking the sorts of societies we might want to live in.

It disturbs me to find abandoned cars in forests, but I find hope in the lichens that reside on them (see Figure 5.3). There's an emergent beauty as they nestle in among the chips of paint and rusting metal (see Figure 5.4). And they occupy not just old cars but gravestones, footpaths, letterboxes, pretty much any surface where they're allowed to live, undisturbed.

Figure 5.3: Lichens colonize metal and paint on an old, abandoned car door

Source: Alison Pouliot

Figure 5.4: A crustose lichen community adheres tightly to rock

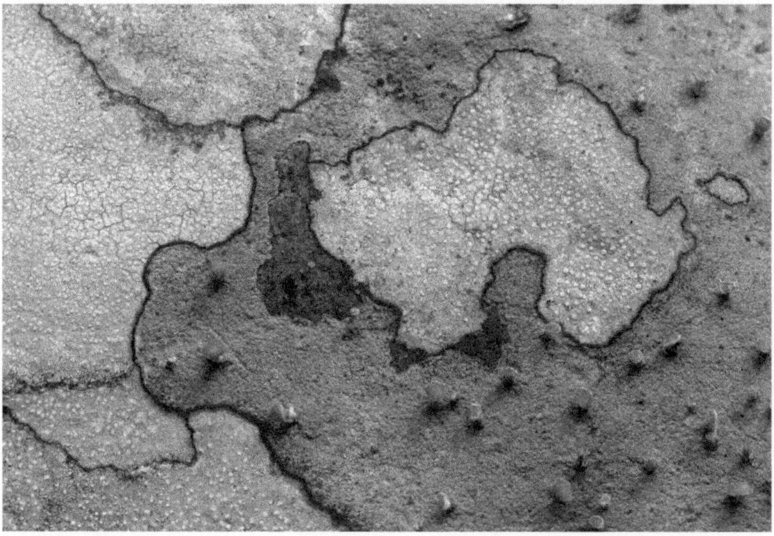

Source: Alison Pouliot

Fungi upend notions of beauty. They challenge us to ask how it might look different: to reconsider beauty in the essential processes of decay and transformation. Lichens create an enigmatic charm as old cars and gravestones slowly metamorphose from lifeless entities to living ecosystems. They offer another aesthetic as human-made structures dissolve and disintegrate. Headstones become hybrids of human history, works of art, and novel microhabitats. Time expressed through a lichenized landscape reminds us that flux and change are the norm.

Maria Brænder and the AMETHYST DECEIVERS: performance and sound artists

AMETHYST DECEIVERS (named after the mushroom *Laccaria amethystina*) is a performance collective (see Figure 5.5) who aim to radically 'fungify' their works, which I initiated as part of my artistic research at Medical Museion in Copenhagen. The AMETHYST DECEIVERS seek to make their performance material with fungi and informed by fungal perspectives. Appearing as otherworldly fungi/human hybrids, AMETHYST DECEIVERS play on self-invented fungi instruments. They also electronically sample live (and pre-recorded) 'fungi vocalizations' in cyclical rituals inspired by four-stage fungi life cycles: spore, germ, hypha, and mature mycelium. Overall, the AMETHYST DECEIVERS strive for a creative process informed by practical research of mycology, and thus embodied experiences and intimate encounters are reflected in modes of collaboration and expression. While honouring the distinctness of its members, composing and improvising are thought of as an ongoing process similar to a mycorrhizal network known to connect the roots of trees in forests, facilitating inter-tree communication (Simard, 2018), rather than clearly delimited individual input.

While the AMETHYST DECEIVERS playfully devote themselves to explorations (bordering on the obsessive) of mycology, they also interrogate how we humans have complex, mingled feelings towards the strangeness of our entanglement with other species like fungi. In the sound excerpt provided (see Figure 5.6 to scan the QR code and listen), the AMETHYST DECEIVERS jam in a response to the contrarian split of emotions they harness towards human/fungi enmeshment. They deliver an expressionistic response to ventilate the body horrors associated with the daily breathing of fungal spores and contemplate healthy fungal lives inside our bodies and on our skin.

Bodily multispecies co-existence often goes unnoticed; however, the AMETHYST DECEIVERS have discovered that a deeper examination of this intricate entanglement can evoke sensations and emotions ranging

Figure 5.5: AMETHYST DECEIVERS, 2025. Current members: Claus Otto, Maria Brænder, and Stephen McEvoy. Costume design in collaboration with Charlotte Bodil Hermansen.

Source: Maria Brænder in collaboration with Sara Valle Rocha and Anders Svanborg Kjeldsen

Figure 5.6: QR code: The sound excerpt is from a ritual, which was part of the exhibition 'The world is in you' (at Kunsthal Charlottenborg, 2022). The ritual was a live event, and so the excerpt is an unedited recording of this occurrence.

from lust to disgust. In their context, disgust may serve as a cultural and neurological response aimed at expelling alien substances from our bodies for self-protection.

The group's lustful desires may be incited as a consequence of an intense fascination and exhilaration directed towards another strange (Queer) species. In view of this, 'lust' pertains more to an enthusiasm for the indescribable essence of existence rather than an erotic sentiment. The AMETHYST DECEIVERS explore the spectrum of the previously described embodied emotions and perceived imaginings through their improvisations.

What's on the collective's agenda is not to make neat, easily read performances that illustrate scientific case studies. Rather, the AMETHYST DECEIVERS are concerned with the Queer potential of mycology. The nonbinary, cryptic, and subversive biological nature of fungi has previously been rendered perverse and unworthy of formal investigation in Western cultures (Kaishian and Djoulakian, 2020). Currently, there's a soaring interest in this field of studies, but the underlying historical resistance towards the queerness inherent in mycology should not sink into oblivion. Rather than traditional science communication that seeks to coolly distance the self from the subject under scrutiny, the AMETHYST DECEIVERS aim to shed light on the uncomfortable, affective, and complex relational aspect we humans always already share with other species. Polemically, pushing it to the extreme, the AMETHYST DECEIVERS want a queered science communication that does not forget the intricate, troublesome, and lively personal (interspecies) relationality.

AMETHYST DECEIVERS have previously performed at Kunsthal Charlottenborg and Medical Museion, between basement, gallery space, pharmacy garden, and anatomical theatre.

On generative interferences: an interlude

This text is an interlude situated in between the voices and works of my three invited interlocutors and a commissioned cartoon by my graphic collaborator Sofie. It is, admittedly, an odd and difficult text to write. As an interlude, this text is neither this nor that – an intermediate text that aims to mediate between the two main parts of the chapter. For Paul Basu, *inbetweenness* marks a relational space of networks and conjunctions but also potentially a void or limbo (Basu, 2017). However, as Basu notes, this space of mixture and hybridity also invites us to be troubled and perplexed as an *interrogatory* space, borrowing the term from Homi Bhabha (Bhabha, 2004; Basu, 2017). In this chapter, I have invited a poet, a photographer, and a performance artist to share their perspectives and practices with us and participate in our exploration of science communication as sustainability. I do not see it as my task here to explicate or interpret their practices, works, and thoughts. Rather I want to embrace the inbetweeness of this relational space and explore how their work might interfere with and potentially trouble our thinking of arts-based approaches to sustainability communication. Within the relational space of this text, I seek to invite interferences that may amplify the hybrid ways art and science are entangled, drawing on the works of my interlocutors previously presented and through personal conversations with each of them.

In communication, interference (or noise as it is sometimes conceptualized) entails the modification, alteration, or disruption of a signal or message. Why then would we want to invite interference? As Warren Weaver notes in his remarks on the mathematical model of communication, while noise increases uncertainty, it also increases the information (Shannon and Weaver, 1998). Yet the additional information is undesirable in Weaver's account, and the crucial question for communication studies is how we can minimize noise and secure a clear transmission of the message. Yet clarity can itself be highly selective indeed, covering up more than it reveals (Grünfeld, 2020). As we emphasize in the beginning of this book, when we conceptualize a new path for sustainability communication, Marshall McLuhan's famous dictum 'the medium is the massage' proves crucial (McLuhan and Fiore, 1967). Indeed, we must consider that messages are not travelling untouched by their vehicle of transportation. Rather, we need to abstain from thinking communication within such dichotomies.

When we consider *science communication as sustainability* as a form of communication that is not just about sustainability but embodies or performs it (see Chapter 1), we are already interfering with previous versions of sustainability communication, be they focused on communicating *of*, *about*, or *for* sustainability (Fischer et al, 2016). And yet we easily get stuck within our usual vehicles of communication trying to shape our messages clearly and disinterestedly. Thus, we need to reiteratively invite in interferences, even if (or perhaps especially when) they challenge our communication pathways and render our work/life more difficult. A central point I want to make is precisely that interferences can be generative for our thinking and doing sustainability communication differently.

In physics, interference means the effect of the combination of two or more waves moving on intersecting or coincident paths (Gregersen, 2025). Imagine throwing two rocks in a lake at the exact same time. Not too close nor too distant to each other. Waves rising and spreading from their impact might overlap to make an interference that amplifies them. Now imagine dropping art and science into the same pool and see what happens. I acknowledge that this image obscures the heterogeneous nature of both art and science and may sound like I'm reducing them to abstract entities, but just imagine how art and science may amplify each other as their waves coincide. The question is how to make them coincide to amplify rather than annul each other.

Likewise, as an act of hospitality, the invitation can also amplify or merge bodies, but it also involves the risk of rejection. As Jacques Derrida argues, ultimately hospitality calls for us to offer ourselves without distinction – to allow oneself to be overtaken by the other (Shepherd, 2014; Derrida, 2023). A crucial question for us then as science communication researchers and practitioners is how we welcome and respond to other perspectives. How are we affected when art and science merge in creative and unexpected ways? In the following, I want to explore how the works and practices of my interlocutors amplify perplexing questions for us to consider as we develop science communication as sustainability.

'Sustainability' is a cosmopolitan concept traversing multifarious discourses from sustainable development goals and policy documents to corporatization and greenwashing. It is also a concept that is hard to love (for precisely those reasons). In our conversations, my interlocutors also expressed a reluctance towards the term, preferring to talk about metabolism, ecology, and climate change. Indeed, Alison prompts us to go beyond sustainability and think about resilience. And as Adam bluntly put it in our conversations; 'sustainability' seems to be a buzzword that we need to treat with caution. In his work, Adam foregrounds metabolism as a conceptual lens for understanding sustainability, emphasizing the competing and contradictory exchanges of breakdown and buildup that challenge reductive ideas of what counts as sustainable or unsustainable.

Yet Maria, Alison, and Adam also seem to share a sense of hope grounded in the transformative powers of metabolic processes. In Adam's combination of poetry and laboratory science, his own metabolism already written by environmental exposures and staged extinction events fuels his writing (Dickinson, 2018). In Maria's work, the metabolic wizardry of fungi transports her to different modes of bodily multispecies co-existences. And in Alison's work, lichens breaking down cars bring a sense of optimism for the future. Although, lichens are not going to save the world – or as she puts it: 'They're not going to eat those cars in any realistic time frame' – they make her think differently, reframing time and space and granting her a sense of hope and aesthetic inspiration amid pollution in the Australian bush.

I think their responses to the notion of sustainability and ways of grounding their practices in concrete exchanges with the more-than-human amplify the need for us to seriously consider how the notion shapes our research and communication practices. While the concept of sustainability seems to be taken for granted within much sustainability communication research referring to the initial Brundtland Report (see, for example, Godemann and Michelsen, 2011), we might need to consider how useful the notion is or how to make it more useful. As we argue in the book (see Chapter 1), sustainability as a concept needs to be expanded beyond its anthropocentric confines. This, however, calls for cultivating a different mindset. If we want to save the concept of sustainability from being reduced to resource management, corporatization, and greenwashing, we need to attend to the more-than-human.

The more-than-human figures prominently in Adam's poetry, in Alison's photographs, and in Maria's performance art. And for good reasons as they – the more-than-humans – reconfigure our experience of togetherness (Maria), how we read ourselves and the environments within us (Adam), and how we fundamentally understand life (Alison). While the figures of the more-than-human in their practices may appear micro as they literally attend to microorganisms such as gut microbes, fungi, and lichens, these protagonists also point outwards to bigger issues. For example, in Maria's practice of cultivating pink oyster mushrooms in old discarded medical books in her home office during the pandemic shutdown of Copenhagen, she found comfort knowing that we are not alone and that other organisms can aid us (as we strive to aid them) in the current sustainability crisis. This sense of complicated inter-species togetherness interfered in her everyday life, pulling her out of her comfort zone and causing existential nausea but also granting her a different (potentially less human-centric) perspective (Brænder, 2021; Grünfeld and Pedersen, forthcoming). This tension between disgust and comfort (even lust as she expresses it) or between horror and kinship resonates with the staged extinction events that Adam explores through his lab-based poetry. While it is easy to be overtaken by horror and despair as we encounter the more-than-human chemicals occupying our bodies, we can also find a kinship with the intraterrestrial microbes that survive extinction events and live inside us as fellow travellers. Or we may find kinship in the unruly renegades of the forest subterrain, if we reach down to earth and look for our fungal allies (Pouliot, 2023).

Perhaps we can understand this shared focus on more-than-human kinship as iterations of Donna Haraway's famous slogan: 'Make Kin Not Babies!' As she argues: 'I think babies should be rare, nurtured, and precious; and kin should be abundant, unexpected, enduring, and precious' (Haraway, 2015, p 163 n 15). But where do we search for kinship? While the body plays a central role in the reconfiguration of togetherness, this togetherness can be sought in the great outdoors (Alison), in the reading of our intraterrestial travel companions (Adam), or in the mesh of sites, people, and more-than human others in the making of a performance (Maria). A provocative question for science communication as sustainability to consider then is: how do we cultivate more-than-human kinship?

Here my interlocutors can guide us a bit along the way to abandon ideals of distanced objectivity and disembodied knowledge and seek out embodied, immersive, and transformative approaches. For example, in his literary approach Adam unfolds an unconventional form of science communication that merges the literary and the scientific to offer alternate ways of approaching the abstract or invisible biological processes in intimate, cultural narratives. When he thinks and writes about pollution, microbes, and metabolism, he tries to enact or inhabit it instead of just writing about it. In his writings, he tries to create an affective experience that invites people to think for themselves about the significance of the issues. He resists the idea that science communication simply translates science to the public; rather, he sees it as an immersive intervention with the potential to offer opportunities for reframing and reconceptualizing key cultural and technological questions.

In a similar vein, Maria emphasizes that her performances should not be perceived as translations (one-to-one) of science but rather as a queering of science communication. Through performance art, Maria together with the rest of the AMETHYST DECEIVERS, Claus Otto and Stephen McEvoy, develop a Queer science communication that embrace the intricate, troublesome, and lively personal interspecies relationality through immersive aesthetic encounters. While this may sound troubling to some, as Alison notes, it is usually through aesthetics that people respond to fungi and only afterwards become interested in ecology.

Like Maria, Alison merges the science and aesthetics of fungi in her work, albeit in radically different ways, to emphasize how they provide creative and imaginative opportunities for thinking differently about the world. For Alison, the forest provides a literal site for thinking-with as she spends extended hours immersed in a practice of *deep hanging around* – writing most of her texts on-site rather than behind the desk. Each of my interlocutors in their distinct approaches trouble what we usually perceive as the core element of science communication – the scientific content – and merge it in embodied, immersive, affective, and transformative experiences.

What then can we – as science communicators and researchers – learn from their practices and perspectives? And more specifically, how can they inform our development of arts-based approaches to science communication as sustainability? As we propose in this book, we need arts-based approaches to science communication to develop other modes of engagement. In our suggestion to terroirize science communication (see Chapter 2), we consider how arts-based approaches may help us land on an Anthropocene Earth shaped by complexities and catastrophes (borrowing the phrasing from Weibel and Latour (2020)). So it might be tempting to ask my interlocutors for advice on how to develop other approaches to science communication and generalize their approaches as tools or methods. Indeed, in our conversations, I also asked each of them what we could learn from their approaches.

Yet this perhaps naïvely formed question ignited unanticipated interferences. For example, Maria responded quite strongly to the question, emphasizing the precarity of being an artist and expressing a well-founded worry about how her livelihood might be affected if science communicators simply appropriate her work as a toolbox. She further expresses her concern about diluting or diminishing artistic practice and study. As she notes, while you would not imagine being able to replace a scientist with someone untrained, why do you think you can do artistic research without the artist?

I think Maria's worry is pertinent and informative when we discuss arts-based approaches in this book. Both because it points to the danger of instrumentalizing art and because it emphasizes the need to include artists in arts-based science communication. Indeed, collaboration and transdisciplinarity appear to be key components in the practices of all my interlocutors. In Maria's performance art, collaboration and openness across differences (artistic formats, disciplines, species) are key features. And as Alison emphasizes, working across species and disciplines can be generative in ways that data points just can't be. In Adam's metabolic poetics, close collaboration with lab researchers invokes critical questions and produces art that shifts the frame of signification, inviting us to think differently about things we didn't think about before. So if we insist on specifying precisely what we might learn from the practices of Adam, Alison, and Maria, we should not look for specific protocols or tools but rather a set of open-ended directions.

Their perspectives prompt us to think about what we mean when we consider arts-based approaches in this book. Arts-based methods have been heralded as an important way forward for science communication (Takach, 2016). The multiple affordances of arts-based methods have been emphasized by other researchers previously as including: a methodological redefinition of research forms and representations (Knowles and Cole, 2008); an epistemological pluralism that recognizes multiple ways of knowing including sensory, kinaesthetic, and imaginative (Gerber et al, 2012); a transgression of art and science that underscores how knowing is more than just knowing *about* (Leavy, 2011); and a reflexive situating of the researcher as part of the enquiry (Sullivan, 2010). Yet arts-based approaches also sit in an odd position between art and science. As Patricia Leavy defines it, arts-based enquiry is not about making art for art's sake but for it to remain a scholarly enquiry guided by defined research questions (Leavy, 2011). Likewise, we could consider arts-based approaches to science communication as guided by the logics of the field and only utilizing or borrowing (or worse: appropriating) methods from artistic research and the arts more broadly.

The risk of this approach, however, is that it remains bounded within the logics of disciplinary scholarship and might end up subordinating itself to a field within which it does not fit. Especially, if we try to challenge the field of science communication to develop performative and embedded modes of communication and research, this path may be futile. Here I believe that Adam, Alison, and Maria may point us in a different and more generative direction that allows the interferences between science and art to amplify each other. But it depends on whether we are ready to be overtaken by a different approach to science communication that embraces personal, embodied, affective, and transformative practices. A paradigm shift perhaps that calls for us to turn to art not for art's sake but neither for the utility of its methods but rather as a generative interference that helps us develop science communication by other means. A turn that calls for us to develop *transdisciplinary collaborations* (move beyond your own discipline) that *diversify* our kin (work with someone you don't know and include other species), *experiment* (in an artistic sense), *speculate* (without being afraid of the uncertain and unknown), and *play* (without curiosity and humour, we are doomed).

As I end this interlude, I am pleased to say that this chapter embodies this call. While I knew Adam, Alison, and Maria beforehand, inviting Sofie – the cartoonist – was not just an opportunity to work together with her for the first time but also an invitation into uncharted terrain. As part of the chapter, we commissioned Sofie to make a comic strip based on the conversations I had with Adam, Alison, and Maria. This was far from an easy task. It called for experimenting, speculating, and playing as Sofie developed the comic in continuous dialogue with myself and our interlocutors. During our conversations, Sofie was present with her notebook, sketching and asking follow-up questions. After the conversations, Sofie began looking for themes that would fit the comic format and noticed that the interlocutors were not necessarily into the idea that their works should be seen as prescriptive. Rather, their practices and works felt more like invitations.

In the following comic entitled *Invitation*, Sofie invites us to perceive their work and practices through the lens of a cartoonist cultivating the subtle tension between humour and seriousness that she encountered across the conversations. Perhaps this shift of perception – an interference in itself – is also a call for us as science communication researchers and communicators to never stop testing other formats to see how they might reorganize our thinking, researching, and communicating.

Sofie Louise Dam: *Invitation*

YOU GO TO A FOREST.
YOU STUMBLE UPON A DUMPED CAR.
THERE'S A CAMERA IN YOUR BAG.
THROUGH THE VIEWFINDER YOU BEGIN TO NOTICE THINGS. YOU NOTICE THE CAR IS COLONIZED BY LICHEN.
YOU ADJUST YOUR SPEED AND SCALE TO THE LICHEN (IN FACT, IT HAPPENS AUTOMATICALLY AS YOU OBSERVE).
METAL, PAINT, PLASTIC, GLASS.
ALL WILL BE GONE IN TIME,
ALTHOUGH YOU WON'T LIVE TO SEE IT.

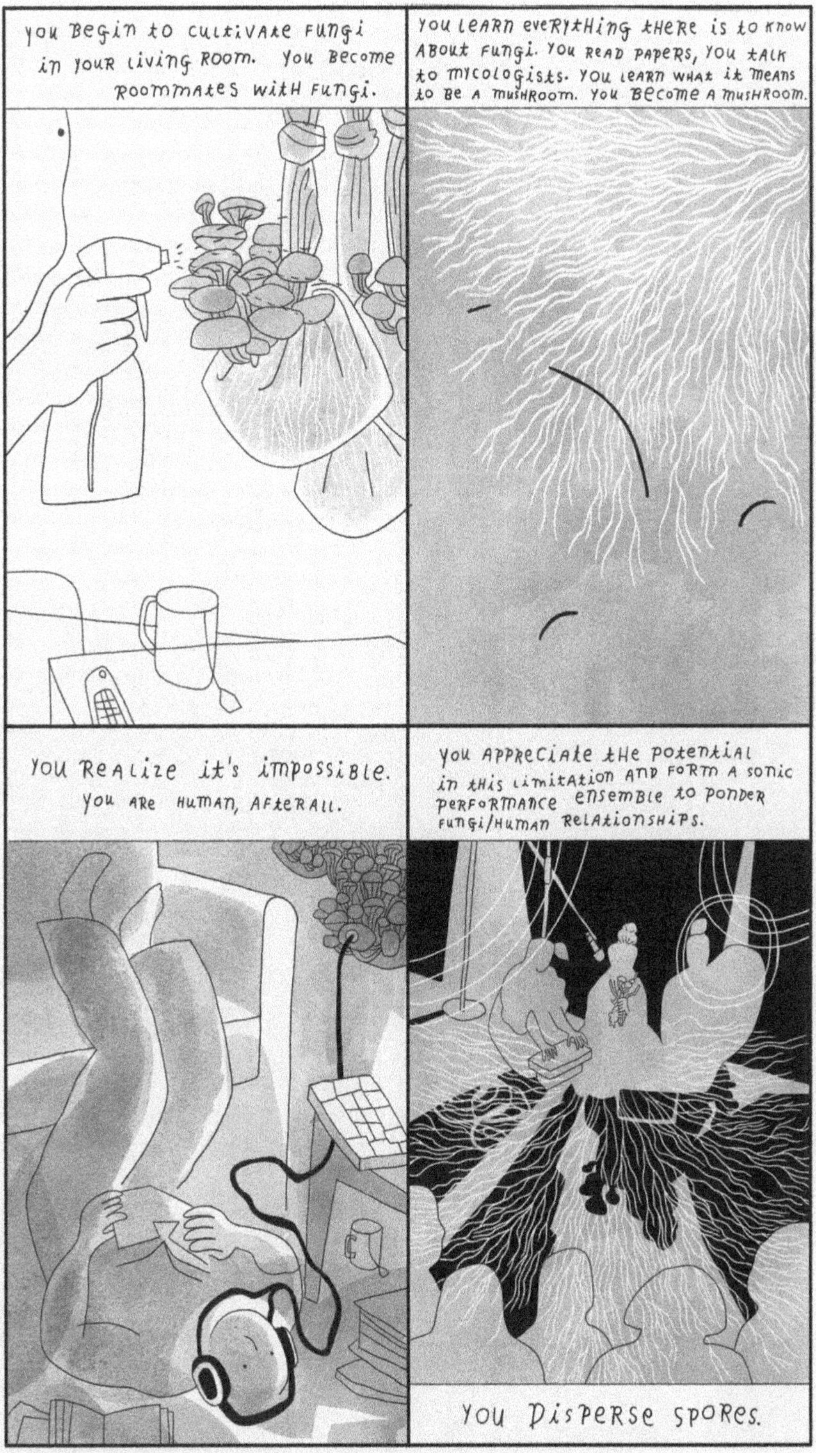
you Begin to cultivate fungi in your living room. You Become roommates with fungi.
You learn everything there is to know about fungi. You read papers, you talk to mycologists. You learn what it means to be a mushroom. You Become a mushroom.
You realize it's impossible. You are human, afterall.
You appreciate the potential in this limitation and form a sonic performance ensemble to ponder fungi/human relationships.
You disperse spores.

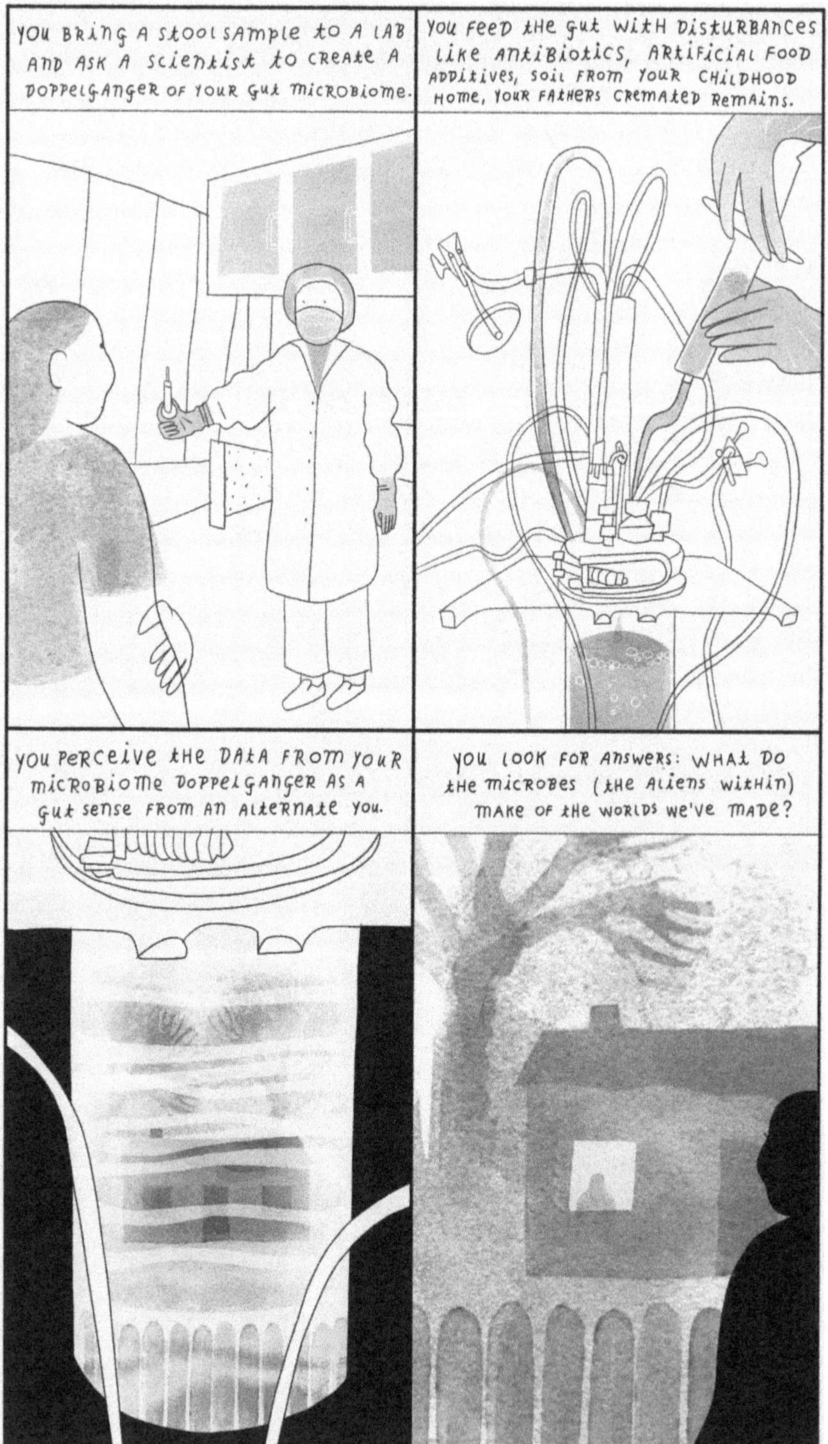
YOU BRING A STOOL SAMPLE TO A LAB AND ASK A SCIENTIST TO CREATE A DOPPELGANGER OF YOUR GUT MICROBIOME.
YOU FEED THE GUT WITH DISTURBANCES LIKE ANTIBIOTICS, ARTIFICIAL FOOD ADDITIVES, SOIL FROM YOUR CHILDHOOD HOME, YOUR FATHERS CREMATED REMAINS.
YOU PERCEIVE THE DATA FROM YOUR MICROBIOME DOPPELGANGER AS A GUT SENSE FROM AN ALTERNATE YOU.
YOU LOOK FOR ANSWERS: WHAT DO THE MICROBES (THE ALIENS WITHIN) MAKE OF THE WORLDS WE'VE MADE?

YOU SIT DOWN AND GET OUT A PEN. YOU WARM UP BY DRAWING YOURSELF IN THE CHAIR BELOW. A STICK PERSON WILL DO.
YOU DRAW A NON-HUMAN LIFEFORM IN THE SPACE BELOW.
YOU USE:
LINE
TEXTURE
VALUE
COLOR
(IMAGINE APRICOT)
YOU ADD CARTOON EYES OF YOUR PREFERENCE.
YOU ADD LIMBS (OPTIONAL)
YOU TAKE YOUR NON-HUMAN AVATAR FOR A WALK.
YOU SEEK THEIR ADVICE ON THE WORRY THAT'S ON YOUR MIND. YOU WRITE THEIR REPLY IN THE SPEECH BALLOON.

AND NOW YOU TAKE A MOMENT TO EVALUATE:
WHAT HAPPENED? CHECK ALL THAT APPLY.
HOW DO YOU FEEL. CHECK ALL THAT APPLY.
ATTUNED TO OTHER LIFEFORMS INSIDE/OUTSIDE MYSELF
CHEMICAL
CONVERSED WITH OTHER LIFEFORMS INSIDE OUTSIDE MYSELF
CONFUSED
GOT IDEAS.
CONNECTED
PASSED SOME TIME
OYSTER MUSHROOM
OTHER
OTHER
IF YOU CHOSE OTHER, CALL UP YOUR FRIEND AND TELL THEM ABOUT IT.
THIS COMIC BY SOFIE LOUISE DAM IS INSPIRED BY INTERVIEWS WITH AND THE WORKS OF ADAM DICKINSON, ALISON POULIOT AND MARIA BRÆNDER/AMETHYST DECEIVERS.

References

Basu, P. (2017) 'The inbetweenness of things', in P. Basu (ed) *The Inbetweenness of Things: Materializing Mediation and Movement between Worlds*, Bloomsbury Academic, pp 2–20.

Bhabha, H.K. (2004) *The Location of Culture*, Routledge.

Brænder, M. (2021) 'Dreaming of big noisy mushrooms: A peek into the first stages of a sonic experiment', [blog post], 26 January, Available from: https://www.museion.ku.dk/en/blog/dreaming-of-big-noisy-mushrooms-a-peek-into-the-first-stages-of-a-sonic-experiment/

Derrida, J. (2023) *Hospitality, Volume I* (P.-A. Brault and P. Kamuf (eds); E.S. Burt, trans), University of Chicago Press.

Dickinson, A. (2018) *Anatomic* (1st edn), Coach House Books.

Fischer, D., Lüdecke, G., Godemann, J., Michelsen, G., Newig, J., Rieckmann, M. et al (2016) 'Sustainability communication', in H. Heinrichs, P. Martens, G. Michelsen, and A. Wiek (eds) *Sustainability Science*, Springer Netherlands, pp 139–48.

Gerber, N., Templeton, E., Chilton, G., Liebman, M.C., Manders, E., and Shim, M. (2012) 'Art-based research as a pedagogical approach to studying intersubjectivity in the creative arts therapies', *Journal of Applied Arts & Health*, 3(1): 39–48.

Godemann, J., and Michelsen, G. (2011) 'Sustainability communication: An introduction', in J. Godemann and G. Michelsen (eds) *Sustainability Communication*, Springer Netherlands, pp 3–11.

Gregersen, E. (2025) 'Interference', Britannica, Available from: https://www.britannica.com/science/interference-physics

Grünfeld, M. (2020) *Writing and Thinking in Contemporary Academia: The Poetics of Clarity*, Routledge.

Grünfeld, M., and Pedersen, S.C. (forthcoming) 'Hospitality and its discontents: Towards a metabolic welcoming of fungal relations at the museum', in L. Campos, S. Bouttier, S. Montin, and F.-J. Lapointe (eds) *Microscopic Life in 20th- and 21st-Century Literature and Performance*, Open Book Publishers.

Haraway, D. (2015) 'Anthropocene, Capitalocene, Plantationocene, Chthulucene: Making kin', *Environmental Humanities*, 6(1): 159–65.

Kaishian, P., and Djoulakian, H. (2020) 'The science underground: Mycology as a Queer discipline', *Catalyst: Feminism, Theory, Technoscience*, 6(2).

Knowles, J.G., and Cole, A.L. (2008) 'Arts-informed research', in L. Given (ed) *The SAGE Encyclopedia of Qualitative Research Methods*, SAGE Publications, Available from: https://doi.org/10.4135/9781412963909.n18

Latour, B., and Weibel, P. (eds) (2020) *Critical Zones: The Science and Politics of Landing on Earth*, ZKM Center for Art and Media, MIT Press.

Leavy, P. (2011) *Essentials of Transdisciplinary Research: Using Problem-Centered Methodologies*, Left Coast Press.

McLuhan, M., and Fiore, Q. (1967) *The Medium Is the Massage: An Inventory of Effects*, Random House.

Pouliot, A. (2023) *Underground Lovers: Encounters with Fungi*, Newsouth.

Shannon, C.E., and Weaver, W. (1998) *The Mathematical Theory of Communication*, University of Illinois Press.

Shepherd, A. (2014) *The Gift of the Other: Levinas, Derrida, and a Theology of Hospitality*, James Clarke & Co.

Simard, S.W. (2018) 'Mycorrhizal networks facilitate tree communication, learning, and memory', in F. Baluska, M. Gagliano, and G. Witzany (eds) *Memory and Learning in Plants*, Springer International Publishing, pp 191–213.

Sullivan, G. (2010) *Art Practice as Research: Inquiry in Visual Arts* (2nd edn), Sage Publications.

Takach, G. (2016) *Scripting the Environment: Oil, Democracy and the Sands of Time and Space*, Springer International Publishing.

6

We Need to Make Things Together: Multiplicity, Kindness, and Transparency in Co-curation Processes

Louise Whiteley

Many visions of sustainable futures involve working better together – within existing communities, but also via the formation of new collectives that bring together diverse ways of knowing and valuing. This book argues that science communication as sustainability is 'a radically collectivist endeavour' that involves and affects multiple people including laypersons, researchers, and stakeholders taking shared ownership within inclusive knowledge sharing and decision making to develop new understandings of sustainability and remake other futures (Chapter 2). This is not an easy task. In the present chapter, I argue that 'making things together' and an attention to the interpersonal processes involved can help. Making a shared material object can act as a creative constraint, opening space for differing forms of knowledge, as well as inviting affective and ethical currents to intertwine with cognitive negotiations (see Chapter 4). The constraint of making operates in part by slowing down thinking and talking; holding time for emergence as materials resist and shift (see also Chapters 3 and 8 which engage with the temporalities of imagining and deep listening). Making together also helps to surface questions of resources, equity, and prioritization, as decisions have to be made about concrete matters such as which materials to use, what to title a work, or where to place it.

Many things can be made together – this chapter focuses on interdisciplinary exhibitions as an 'intensifying case' with relevance to other communication formats, with examples drawn from my work at Medical Museion at the University of Copenhagen (www.museion.ku.dk). I begin by introducing

Medical Museion and the centrality of its collections, then unfold the role and handling of multiplicity, discuss exhibition making as an 'experiment' at the intersections of theory and practice, and argue for the value of artistic collaboration. Medical Museion addresses the past, present, and future of medicine and is not specifically focused on sustainability, but these 'moves' – to collect multiplicity, co-produce knowledge, and work with artistic approaches – are also at the core of the call for sustainability science communication articulated in this volume. In addition, sustainability is increasingly present in discussions about the future of health and medicine (see, for example, Brown et al, 2013), and what it means to be a body embedded in environment and in multispecies relations (Rupprecht et al, 2020; Formosinho et al, 2022). Throughout, I therefore also reflect on how our curatorial approach relates to the proposals made for sustainability science communication in this book.

In the second part of the chapter, I move away from discussing exhibition formats and focus on the people making them. In any form of collective work, it is not just knowledge that needs to be combined but knowers that need to gather. And knowers are not just cognitive machines but persons and their relations, histories, dreams, anxieties, quirks, limitations, and resources (see also Chapter 4). Studies of fields ranging from makerspaces (Davies, 2017), high-energy physics (Traweek, 1988), to hippie communities (Berger, 2004), and conflict resolution (Oetzel and Ting-Toomey, 2013) have curiously observed how such groups function, feel, and communicate. I take inspiration from these fields, but my contribution is not primarily analytical. Rather, with my curator hat on, I share a series of suggestive principles for making things together, based in experiences of both flow and trouble in curatorial collectives, which might be useful in other kinds of 'making spaces'. These principles are not intended to be universal – in fact, a grounding principle is that all such principles must be remade (or thrown out) in context. In conclusion, I reflect on the limitations and potential harms that might be invoked by applying these principles unreflectively.

Collecting moves

Medical Museion is situated in a complex of 18th–19th-century buildings in the heart of Copenhagen. The main museum building was constructed in 1787 to house the Royal Academy of Surgeons of Denmark, centred around a spectacular auditorium used originally for dissections and demonstrations, which today hosts public events, PhD defences, performance lectures, and tours. In 1907, a trade exhibition for the Danish medical profession became a collection housed in these buildings, which was later incorporated into the Department of Public Health at the University of Copenhagen and then in 1999 renamed and envisioned as Medical Museion (Söderqvist and Pedersen,

2013). The name is a nod to the ancient Greek notion of a 'mouseion' as a temple of the muses – a place for multi-disciplinary investigations (Arnold et al, 2021). In our version though,[1] the muses are not a cast of female allegories draped in floating fabric but an ever-changing cast of people, practices, and ideas – if one still dominated by Western ways of imagining science, health, and medicine.

A collection is typically seen as a necessary ingredient of a museum, and this museum holds around 100,000 objects, depending on how you count:

> [A]n amputation saw; false teeth recovered from an old hospital laundry; a human arm preserved in slices, colored pins marking out the parts of the whole; a ceramic dissection table; thousands of tissue samples in wax blocks; elegant glass X-ray tubes – a steampunk dream and a cleaner's nightmare; apothecary jars marked POISON; and a plague wagon. And then there are collaborative productions sitting somewhere between object and artwork: an interactive screen showing the artist's skin merged with landscape photography and an MRI of their guts[2] next to a historic microscope used by Hans Christian Gram; a family portrait made of projected microbiome sequences;[3] and a dress woven from the 27,774 pills an average woman with metabolic syndrome would take in ten years.[4] (Whiteley, 2025, p 213)

In her discussion of encounters with colonial objects, Sandra Dudley writes that in museum collections 'we are most likely to find things that "fascinate, awe, shock, irritate, or puzzle"' (Dudley, 2013). Note that Dudley does not use verbs such as inform, persuade, solve, or calm. Medical museum objects also provoke feelings, often powerful ones – a desirous impulse to touch a poison bottle; a sinking feeling when confronted with a cabinet of dental pliers; sadness at the interpolated body in the stirrups; prickling rage for phrenology callipers. Museum objects also invite multiple stories and curiosity about context – who used them, how, when, why, and where? And questions thus about absence – who didn't use them, and what might have happened to their bodies instead? Museum objects are thus natural 'gathering places' for multiple perspectives – and curating multiple objects into an exhibition is an extension of this. Even more questions are raised; possible narratives proliferate. At Medical Museion, we further expand the multiplicity by combining objects from cultural history with artistic works and scientific objects – inviting people from multiple disciplines to help with this combinatorial 'making', and when resources allow, to make new hybrid objects together too. But these combinations or collages are not just lists; they contain 'connections, spiky contradictions, and delicious juxtapositions', differences that open new avenues via their irresolvability (Whiteley, 2025, p 217).

Figure 6.1: An assemblage of perspectives around the theme 'Look inside' in the exhibition *Mind the Gut* at Medical Museion, University of Copenhagen

Source: Morten Skovgaard, copyright Medical Museion

In *Mind the Gut* (2017; Whiteley and Bencard, 2020) one of the displays gathers objects around a shared theme 'Look inside' (Figure 6.1). The famous microbiologist Hans Christian Gram's brass microscope is displayed with some of his original glass microscope slides and contemporary biopsy slides from artist Naja Ryde Ankarfeldt's collaborative work *Landscape Epithelia*,[5] which is itself projected onto a nearby curtain. The projection merges images from Naja's gut biopsy, brain scan, skin, and landscape photography. In the same assemblage, history books showing photographs of Ivan Pavlov's experiments measuring dog's gastric fluid in emotional states sit next to a hyper-realistic silicone torso by Michael Schultz, representing Alexis St Martin (1802–80), who in 1822 received a bullet wound that became a permanent fistula used by army doctor William Beaumont to make seminal discoveries about gastric acid. Multiple perspectives on the act of *looking within* are present here, ethical, epistemic, and individual questions about how to live with the breaching of our bodily boundaries are alive in the space. The exhibition invites curiosity as well as critique – though this ethical unsettling is of course itself an ethical stance.

So museums collect *things*, but they also collect perspectives, stories, and ideas, by temporarily collecting groups of people and weaving their perspectives into the selection, arrangement, and even creation of objects. In the words of Weibel and Latour, museum exhibitions are 'a highly artificial assemblage of objects, installations, people and arguments, which could not reasonably be gathered anywhere else' (2007, p 94). Curation also involves cross-professional teams inside the institution, and often invites outside voices in to consult and contribute, sometimes creating sustained co-curatorial relationships where, for example, scientists, artists, and humanists are all invited to participate in development of the 'total medium' of the exhibition over an extended period of time (Bencard et al, 2019; 2022). For instance, artists might comment on scientific content and critique marketing; and scientists might help develop design choices or visitor surveys – stepping outside their usual disciplinary roles while respectfully acknowledging their naïvety. When exhibits open, the proliferation continues through the often-unpredictable links that visitors make to their own lives, in informal conversation or programmed events, which offer opportunities to augment the collage of perspectives still further.

Why multiplicity?

In her work on how we might communicate with incomprehensible microbial others, Maya Hey draws on the work of John Durham Peters to 'reorient human–microbe communications towards polylogue, where many speak and many hear simultaneously' (2023, p 11). I suggest that this situated concept can be usefully paired with the notion from Robert Grudin (1996) of dialogue as 'reciprocal and strange' (see also Madvig et al, 2022) – whereby multi-perspectival encounters involve accepting incomprehensibility rather than trying to dissolve it, orienting ourselves to better hear others, and welcoming the strangeness that is the very purpose of talking (or making) together. It is tempting to think of a polylogue as more representative than a dialogue – by definition, it involves more than two voices. But those voices were still selected, and framed, and explained, and many were left out; practices of inclusion often shaped as much by practicality as by ideology and anxiety. Collection, curation, and exhibition are never neutral (Greenhill, 1992; Evans et al, 2020), and we should beware 'multiplicity washing'. The writer Chimamanda Ngozi Adichie (2019) said in her TED talk: 'It is impossible to talk about the single story without talking about power.' This is true for the multiple story, the polylogue, the assemblage, and the collage too. So by creaking open the usual library of imaginative resources (see Chapter 3); the usual mixing desk of sounds (see Chapter 8); the usual emotional registers (see Chapter 4); the usual historical cases (see Chapter 7); and inviting in new knowledges and knowers (see Chapter 5), we must not assume that the work is done, or that the oppression of the single narrative is over.

Attempts at bringing multiple knowledges together in an exhibition challenge a long history of seeing museums as instruments of ordering and categorization. Whether collecting closely related butterflies to derive taxonomy, or collecting human remains from people with diseases to bound the 'normal' body, gatherings of things can be used as much to corral complexity as to unfold it. As Nicholas Poole (2014) asks in his essay about engaging complexity in museums:

> Can we take a model that has been optimized over 200 years to create the impression of order and pattern, to deliver 'single stories' of historical progression, and use it as the raw material for a new industry whose primary purpose is to equip everyone with the tools to engage more productively and creatively with complexity?

An 'STS (Science and Technology Studies) sensibility' can help us here, combining an attention to materiality and how it is manipulated to find manageable pattern in complexity, with a keen sense for messiness and contingency (see also Bencard and Whiteley, 2026). According to Law and Lin (2022): 'To approach anything in STS is to attend to how weavings create objects, people, texts, and proper relations – versions of the good and the bad. It is to know that it is possible to discover weavings all the way down.' Exhibitions can materialize this sensibility; assemblies of assemblies (Weibel and Latour, 2007).

Thus far, I have tried to present interdisciplinary, co-curated exhibitions as a great way to explore and exhibit multiple perspectives. But let's step back for a moment and address the assumption that this is a valuable goal. Why exhibit multiplicity, and why invite multiplicity into the process of curation as well as the objects curated? Why unfold hard-to-grasp complexity rather than tell compelling single stories? We and others have given various answers to these questions, which blend normative principles with methodological appeals – and which resonate with the arguments made throughout this volume in relation to sustainable science communication as necessarily multivocal, multisensory, multimodal. The topics we deal with – from mind–gut connections in *Mind the Gut* to soil sustainability in *Fallowland*[6] – belong properly to no one, they are 'constitutionally collages of epistemologies, and as such any attempt to investigate, create, or publicly engage with them should mirror this multiplicity' (Whiteley, 2025, p 215). In other words, the world is this way, so we should engage with it as such. This argument might seem like a naïve kind of scientific realism, but it doesn't imply 'accurate representation' – quite the reverse, it acknowledges the limits of addressing the world as and via a composite of separable elements. The exhibition has both a capacity and thus perhaps an imperative to engage people in assemblages, in weavings: to 'make visible that which is otherwise invisible, to make tangible something intangible' (Basu and Macdonald, 2007, p 6). Below the surface here is a belief that visitors – publics – can cope with and even relish complexity. They live in this world too.

So we want multiplicity, but why not channel the multiplicity of the world through the inquisitive mind of a single curator? An open-minded historian could research clinical, philosophical, or patient perspectives when making an exhibition on transplantation and would do so more efficiently than an awkward co-curatorial group. However, the bearers of those differing perspectives can express their expertise more fluidly, associatively, and aesthetically when they gather together, to work with the medium as well as the message. Inge Daniels (2019) in her book *What Are Exhibitions For?* writes that co-creation brings together 'mutual possibilities and mutual difference … grounded in the particular relational emergence of knowledge between people' (p 207). In the latter part of the chapter, I will explore some elements of this 'relational emergence' and how to support it.

Knowledge-generating 'experiments'

Collecting practices have always generated knowledge, whether in a museum, archive, or biobank (see Tybjerg and Grünfeld, 2023), and exhibition curation can also act as an open-ended methodological experiment in what kinds of knowledge might emerge (Macdonald and Basu, 2007; Bjerregaard, 2020; Arnold et al, 2021, p 59). As discussed earlier, this knowledge might be of the ordering and categorizing kind, or it could be more grounded in association and expansion, both of which can be seen as 'experiments' when we let that concept travel flexibly across disciplinary traditions. Basu and MacDonald place the historical siblings of exhibition and experiment back together, arguing that

> the exhibition is thus conceived as a kind of laboratory, in which, to use the language of actor network theory (Law and Hassard 1999), various 'actants' (visitors, curators, objects, technologies, institutional and architectural spaces, and so forth) are brought into relation with each other with no sure sense of what the result will be. The exhibitions discussed are, it might be said, experiments in meaning-making. (Basu and MacDonald, 2007, pp 2–3)

In their work on transdisciplinary collaborations between neurosciences and social sciences, Fitzgerald and Callard (2015) put forward a related notion of 'experimental entanglements'. Here, the neuroscientific experiment is not treated as a strictly regulated disciplinary machine but as a space for collectively reconstituting ideas about what we could find out. Collaborators aim to impact each others' practices as well as the object of study, starting from their disciplinary positions but through collective making striving not to reinscribe disciplinary boundaries. An exhibition can also invite such 'experimental entanglements', if we make space and time for them to happen (Bencard et al, 2019, p 139).

Seeing curation as experiment echoes the call in this book to enthusiastically explore relations between theory and practice in science communication to relish 'the productive tension between scholarly inquiry and lived experience' as a path to sustainability (see Chapter 1). Beyond using science communication events to gather data, as in citizen science initiatives, we can adaptively investigate the issues we all live and research inside; a form of relational learning (Spangenberg, 2011; Canfield et al, 2020). Exhibitions and other ways of 'making together' create a material as well as discursive laboratory for these investigations. Ian Bogost (2012) proposes the notion of 'philosophical carpentry' – arguing that philosophers can construct artefacts that do philosophical work. Adam Bencard (2019) suggests that we apply this to exhibition making, and in an article written with Martin Grünfeld (Grünfeld et al, 2023) we proposed the notion of 'metabolic carpentry' to explore 'how materially producing metabolic paradoxes might open new avenues for thinking about their concrete consequences' (p 388). Perhaps we could continue the word play and think about 'sustainability carpentry' as making things together that also construct the kinds of thinking, feeling, and acting that the group discovers are needed for sustainability.

Artist collaboration and octopus stools

The carpentry we have developed at Medical Museion often involves bringing artists together with researchers and others to make collaborative work. Rather than pairing one artist with one scientist, we have preferred to set up 'three-legged stools' between art, science, and social science or humanities (which we have more recently situated in relation to the development of the Art, Science, and Technology Studies literature, see Rogers et al, 2021; Hussey et al, 2026; Bencard and Whiteley, 2026). Here the relation between artist and scientist, between art and science, is not predetermined and can range from direct engagement with a particular experiment to aesthetic inspiration or shared ethical reflection. The three legs are rarely equal in their role or influence, and we should not pretend that they are – this is a stool that might not stand up. This is even more so when expanding the co-curatorial team outside of the 'academic' sphere, and when participants hold hybrid roles that shift during the project. This is perhaps an octopus rather than a stool – smart but slippery, flexible and capable of hanging upside down, if rarely evenly balanced.

So why make art-science octopi? Muller et al (2020) describe art-science projects as a transdisciplinary 'third space', with the capacity to 'to connect audiences and stakeholders to researchers in ways that are mutually enhancing' and where 'emerging, empathic and ethical knowing' is produced (p 321).

In both sustainability and biomedical challenges, experiential knowledge from the people who live those challenges is critical to generating sustainable outcomes for real-world contexts; bringing forth knowledge that the makers feel to be missing (Achiam et al, 2021; Verlie, 2022). Experiential knowledge is welcomed and frequently centred in artistic practice and art-science 'third spaces', alongside sensory and material dimensions (Gerber et al, 2012; Damsgaard and Phoenix, 2022; Achiam et al, 2024). As in the sister field of Speculative Design (Dunne and Raby, 2013), making things together in this setting can help to render alternative ontologies tangible (Born and Barry, 2010), enacting other pasts and futures (see Chapter 3; see also Bencard et al, 2019). As Susan Sontag claims in her essay 'On style':

> Art is not only about something: it is something. A work of art is a thing in the world, not just a text or a commentary on the world. … Which is to say that the knowledge we gain through art is an experience of the form or style of knowing something, rather than knowledge of something. (Sontag, 1969, p 30)

In *Pill Machine* (Bencard and Whiteley, 2024),[7] artist Mogens Jacobsen worked with electronics collective Circuit Circus and the co-curatorial team to develop an interactive installation that would invite visitors into the more 'holistic' perspective on health that can come from examining connections between mind and gut. In the 'third space' of our group meetings, we came up with the concept of a diagnostic machine that would enact the exhibition's vision of what might matter for someone's health. Diagnosis was a practice that seemed capacious enough to gather multiple perspectives and invoke some of the sociopolitical questions the team were interested in around who governs medical knowledge. From the curators' perspective:

> Diagnosis can also unlock access to resources, change our legal status, shape how hospitals are organized, affect our identity, and how others see us. It gives a name to something that is often hard to grasp, though names can be both liberating and oppressing. *Mind the Gut* contains many cultures of medicine, evoking many diagnostic systems. *Pill Machine* attempts to gather them all but drops things along the way; a doctor committed to holism but frustrated in clarity … The machine remains silent on whether such a holistic diagnosis is possible, but it prescribes words and medicine nonetheless. (Whiteley and Bencard, 2020, p 111)

The machine here was the thing we would make together, in a thin corridor space that suggested a series of questions that the visitor would answer before receiving a prescription at the end (see Figure 6.2). This format of eight input

Figure 6.2: *Pill Machine* in the exhibition *Mind the Gut* at Medical Museion, University of Copenhagen.

Source: Morten Skovgaard, copyright Medical Museion

devices gave us space for multiplicity: we brainstormed an extensive list and selected and refined the final eight by combining our different perspectives and values. The questions include how sharp your boundaries are with the world; how many antibiotics you take and how much junk food you eat; where do you feel your emotions; and whether you are in love. The installation was beloved by visitors until it was finally decommissioned in 2024, and the artist and curators on the team rescued parts of it to keep in their offices – its questions continuing to voice themselves.

Mirroring content and ethos of outcomes in curatorial processes

Thus far, I have argued for interdisciplinary, co-curated, arts-based, 'made together' exhibitions as an intensifying case for some key visions of this book. As a method for bringing knowledges together in a way that can challenge traditional modes of simplification and categorization and unpick hierarchies of knowledge and practice, drawing lived experience into relational learning for sustainable solutions (Spangenberg, 2011; Canfield et al, 2020). This echoes a story often told about the evolution of science communication: moving from one-way dissemination to 'speaking and listening' formats that were subsequently unpacked as rife with unspoken power relations and powerful silences, and more recently challenged by definitions of science communication as culture, identity, or even terroir (Davies and Horst, 2016; Achiam et al, 2025; Chapter 2 in this volume). When viewed through these more situated lenses, participants in science communication are not always speaking in the same translatable register, and voices are not the only communicative mode that matters. Maybe the poly-*logue* is incomplete without its imbricated poly-*motus* – though etymology insists that we hold the implicit binaries between reason and feeling lightly. With apologies to the classicists, *logos* is word but also action; *motus* is feeling but also movement. Thinking through science communication as culture or terroir also reminds us that people are not just elements of complexity but complex beings themselves.

In this section, I focus in on the individuals making things together rather than the visitors who will come later. In other words, I apply the same wishes we have for arts-based science communication audiences to those who are involved in producing the communication itself. I offer some suggestions for how to facilitate such processes, based in my experiences of both flow and trouble in curatorial collectives. This is from my perspective, and I attempt not to speak for others but am deeply informed by their generosity and insights. So why focus on the makers rather than the audiences?

On the one hand, the two groups bleed into each other. The makers are part of our audience and part of networks with whom they might share the deeper engagement they experience during a co-curatorial process. On the other hand, audiences can be part of the co-curatorial development, from one-off events to more embedded involvement, and thus we want the same values to be inscribed across the process. This can help the team to create a shared sensibility, which can take the pressure off finding shared concepts, knowledge, or moral values. Methodologically, the process of co-creation can also be seen as a kind of 'pilot work' for public activities, where the co-curators represent not just their professional expertise but their lived experience. Work on participatory practices in museums is a vital relation here (Simon, 2010), where the 'making together' described in this chapter is an (arguably rather exclusive) microcosm of larger-scale work aiming to bring diverse local audiences into museum programming. We argue that this microcosm is also interesting in its own right – how knowledges, feelings, values, and priorities are negotiated and related in the 'making together' can be studied as part of interdisciplinarity or curatorial research as well as a potential source of knowledge about the topic at stake (see, for example, Basu and Macdonald, 2007).

Suggestive principles for making together

The suggestive principles in this section are largely about modes of interacting and being together, and I hope they might be useful as inspiration, provocation, or just 'extra flavour' in a range of communication activities. They are also personal, and here I follow the other authors in this volume who bring their own experiences into play in their writing. Just as we see scientists not just as knowledge producers but 'as citizens, as activists, as artists, as parents' (see Chapter 4), we also see ourselves as fully present in our work and in our heartfelt attempts to co-weave less heartbroken futures together.

Making things together requires communication; to make sense (or perhaps non-sense) of multiple ways of knowing. Scientists and philosophers have long asked how different kinds of explanation, on different scales, and in different languages, can be made to work together. The dream of translation between disciplines is only possible in limited cases – which perhaps shouldn't surprise us in light of the creative conundrums of literary translation, before we even approach the languages of chemistry, biology, and sociology. So what then? Merge related terms? Create pragmatic bridges between incompatible terms to allow parallel investigations? Social studies of scientific collectives offer deep insights into how these strategies unfold in practice, amid webs of people, technologies, experimental materials, infrastructures, career constraints, and more, finding that people continue to collaborate even when such philosophical puzzles cannot be solved. For instance, as captured by Star and Griesemer's (1989) famous 'boundary concepts' that can hold different meanings but still guide collective investigation, or in classic observations from medical sociology about people's nuanced ability to hold together apparently incompatible beliefs or deploy them in ways that might be technically 'illogical' but are deeply 'rational' in the context of their lives.

In my curatorial work, I start from the assumption that translation is always imperfect – and that failures of understanding can be the starting point for creative collaboration, not just a roadblock to it. To unlock this potential, it can be helpful to ask which norms make us anxious to *mis*-understand each others' work – and whether some of the tangles we find ourselves in come from the mismatch of those norms with collaborative creative practices. From this starting point, explicitly mushy boundary concepts, 'holding spaces', gaps, and playful euphemisms can stabilize a collective in order that they can proceed to make something together while preserving the rich resource of concepts between them. This is fragile and requires care-ful work; organizational, emotional, institutional labour that demands transparent negotiations about its distribution, reward, and recognition. In Callard and Fitzgerald (2015)'s reflections on their own practice creating experimental entanglements between the neurosciences and the social sciences, they position interdisciplinarity as 'an affectively fuzzy domain, and much of the work of collaboration as learning to live though that fuzz' (p 112).

To return to the *Pill Machine*, we did not try to align or translate notions of diagnosis between the artists, historian, electrical engineers, curator, and designer involved. Rather we inverted this demand and relished the looseness of the term *diagnosis* while getting very specific about which diagnostic *questions* could be asked from the tangle of our different interests. The machine – the thing – provided the frame for this, along with the shared vision of inviting visitors to experience an interaction that would both remind them of their own diagnostic experiences and be completely fantastical at the same time. In another project, we wanted to produce a 'spiky experience' for visitors. This metaphorical expression was used to denote a feeling of non-smoothness that we all felt we agreed upon, but without having to reconcile our different ways of conceptualizing what should happen to visitors' knowledge, feeling, or action.

Another gathering space for groups making together can be ongoing conversation about values, origins, hopes, and fears. We often invite people to begin a collaboration by talking about where they come from; what brought them to their practice; what brings them joy and frustration. People are always present with their values, driven by what matters to them, but this is often unspoken, hidden behind the rubric of *disciplinary* values, or erupting in moments of frustration. By inviting people to bring those values forth and celebrate the texture they bring to knowledge making, I argue that we can create threads of understanding that better hold up when tensions arise. In her parenting podcast, Dr Becky Kennedy finishes every episode by asking the listeners to place their hand on their heart and say to themselves: 'Even as I am having a hard time on the outside ... I remain good inside.' It would be a very special circumstance in which I would invite a co-curatorial group to engage in this potentially 'cringey' activity, but the spirit of it is present when I reiterate a group's shared commitment to making something meaningful, articulate how our institutional responsibilities can constrain our ability to 'bend' to others' priorities, and express trust in the group's ability to navigate tension and disagreement.

Another activity that can be useful is to brainstorm hopes and fears for a project, or, in more dramatic language, dreams and nightmares. The facilitator can set the tone by presenting honest and sometimes playful items to bound the list, and to show that naming these bounds can help to put them in their proper place rather than rendering them overwhelming. This work also takes us towards the space of imagination (see also Chapter 3), within the microcosm of a team making together, loosening the bounds of ideation as when one co-curator shared a dream of flooding the basement of the museum. This was not allowed, but it become an anchoring point for the group's imagination, a reminder to push further, feel *deeper* into what we might hope for a visitor's experience. Returning to a dreams and nightmares brainstorm can also help the group see how they have evolved their conceptions of what could be and what should not be. In one co-curatorial process, we expanded this activity by drawing on a participant's experience with Augusto Boal's Theatre of the Oppressed tradition (Boal and McBride, 2014). One person stood in the centre and spoke 'as the exhibition', 'as the room', or 'as the laboratory mouse' about what *it* hoped and feared. Centring other aspects of the assemblage was not intended to reify anthropomorphism but to help the group displace their feelings around what we were making together and then to experiment with speaking outside their expertise. Tentatively trying to voice other perspectives and correct others' 'mistakes' through playful augmentation.

All these techniques are an opportunity for the group to discuss (a) immovable principles – whether these come from the institution's expectations, an individual member's ethical commitments, or practical constraints; (b) where they can accept/enjoy multiplicity; and (c) where the limits of that multiplicity lie – for example in choosing a title, responding to a media enquiry, or cutting three objects from the exhibition list. Here, a connection emerges to sustainability – where we are challenged with the feelings of immovable constraints but must imagine multiple futures regardless. Accepting misunderstanding and welcoming multiplicity is not about saying everything and avoiding making decisions but about expanding the space and time in which difference can be allowed to entangle. This space and time comes at a cost – but whatever the scale of the project, it can be discussed openly when selections need to be made and why: paying attention to the group's 'fuzzy affects' and asking what they tell us (Callard and Fitzgerald, 2015). For example, when creating visitor feedback cards for an exhibition, strong feelings that an artwork should not be excerpted on the back of the card told us (a) that the artist wished that not to happen, which according to group principles trumped other priorities; and (b) initiated a discussion about the potential violence we might do to the artworks within our hybrid practice by treating them as mere 'aesthetic elements'. This is a particularly important cluster of issues to bear in mind when working with artists and artistic approaches: art has often been instrumentalized by and for science communication, and its practitioners are understandably wary of this occurring (see also Chapter 5).

This example brings me to another suggestive principle – make power, resource, and authority a transparent discussion. In a group making together, there are likely to be members with differing power in the institution (and also different duties and responsibilities to that institution); of different levels of seniority; and from professions with differing social status and financial reward. Permanent employees of a university live in a different relation to their working hours than freelance artists or students. Acknowledging this can help the group to calibrate what the members can expect of each other, explaining what commitment and priority mean from their positions, and creating an awareness of points where we might need to 'stand up' for each other. However, privacy and legal confidentiality must of course be respected, and checking in with individual team members outside the group setting can help to calibrate the discussion. One might worry at this point about this being 'all talk' – as Mieke Bal notes, 'openly acknowledging the authorial role of the curator does not necessarily challenge the curator's authority' (2007, in Basu and MacDonald, 2007) – promising to change is not enough.

In engaging challenging topics such as disagreement, fears, ethical principles, and power and authority, I find it helpful to talk about kindness. As I write this chapter, the question of *kindness* is bubbling up in a range of places – from calls for academic research culture to be kinder[8] to LGBTQIA+ memes and merch calling for others to 'be kind'. These calls are often inadvertently (and sometimes carelessly) grounded in ancient ethical and spiritual traditions, which should remind us that they also grow out of responses to deep suffering, and the difficulty of grappling with the *lack* of kindness that devastates on a global scale. But the popularity of this concept also makes it an accessible way to talk about feelings and ways of interacting with each other – if space is also held for discussing the potentially patronising impacts of 'kindness' distributed as largesse and the dangers of 'kindwashing' in the face of wider structural inequalities and discrimination.

Conclusion

In the chapter, I have mentioned 'multiplicity washing' and 'kindwashing' as dangers. We need to be wary of terms such as multiplicity, kindness, holism, and inclusivity that have a comforting ring to them, at least to those like myself educated in a particular politics of science communication and sustainability. In groups making exhibitions – or sustainable futures – we need to collectively articulate what values those words carry, and which structural injustices we might be turning a blind eye to by nourishing them within smaller groups. But on the other hand, cynicism also has a siren call, and it is easy to neglect the wellbeing of those trying to make a difference by pointing to their privilege. Making things together can help to hold both dangers in tension, and in parallel. I also cautioned against an assumption that multiplicity is automatically democratic: even polylogues are selected, chosen to present particular subsets of perspectives, and the selection itself forms a new perspective. No story is neutral; no narrative format is value free.

Implications for science communication as sustainability

This book calls for deep sustainability as an approach that not only addresses systemic issues but questions underlying cultural, philosophical, and epistemic relations (see Chapter 1). In this chapter, I hope to have presented 'making together', grounded in the intensifying case of interdisciplinary exhibitions, as a method for holding such questioning while also participating in creating sustainable futures. To end, I offer the following suggestive principles for 'making things together' towards sustainable futures:

- don't expect to find a singular 'shared language', vision, or conceptual translation;
- instead centre the group's ethos and values, as an always-evolving discussion;
- welcome and relish multiplicity, but be clear about when it needs to be collapsed;
- invite participants to bring their hopes and fears into the project – but with an attention to issues of privacy and safety;
- talk about power, resources, and authority openly; make the limits clear and explore collectively within them;
- be kind and talk critically about kindness.

The suggestive principles need to be made concrete in each specific context and may not be possible or helpful. Perhaps then, the overarching principle is to dare to consider, discuss, and foreground these matters, which also involves sitting with their situatedness and ethical instability.

Notes

1 I use collective pronouns to refer to 'our' work at Medical Museion, and to acknowledge the collective that shapes a lot of the thinking presented here, but I also want to emphasize that the 'we' in question is diverse and evolving, and I do not speak for all members of staff past or present.

2 *Landscape Epithelia*, Naja Ryde Ankarfeldt, 2017.

3 *Micro-Families*, produced in collaboration between Medical Museion and scientist-artist François-Joseph Lapointe, additional sequencing and modelling by Daniel Ortega, Santiago Ortiz, and Virginie Lemieux-Labonté.

4 *Femme Vitale*. Pharmacopoeia, 2013.

5 *Landscape Epithelia*. Naja Ryde Ankarfeldt, 2017.

6 https://www.museion.ku.dk/en/events/fallowland/

7 *Pill Machine*. Created in collaboration with artist and co-curator Mogens Jacobsen and Circuit Circus, 2017.

8 https://uniavisen.dk/en/new-initiative-research-culture-needs-kindness-as-a-core-value/

References

Achiam, M., Dillon, J., and Glackin, M. (eds) (2021) *Addressing Wicked Problems through Science Education: The Role of Out-of-School Experiences*, Springer.

Achiam, M., Vitting-Seerup, S., Whiteley, L., and Dam, S.L. (2024) 'Creating resonance with arts-based approaches to sustainability science communication', *Journal of Science Communication*, 23(6): N01.

Achiam, M., Grünfeld, M., Vitting-Seerup, S., Thorek Jensen, J., and Whiteley, L. (2025) 'Towards a terroir approach to science communication and its evidencing', *Journal of Science Communication*, 24(7): Y01.

Adichie, C.M. (2019) 'The danger of a single story', TED Talk, Available from: https://www.ted.com/talks/chimamanda_ngozi_adichie_the_danger_of_a_single_story

Arnold, K., Bencard, A., Tybjerg, K., and Whiteley, L. (2021) 'Museum as academy: Research practices at Copenhagen's Medical Museion', *The Garage Journal: Studies in Art, Museums & Culture*, 3: 52–81.

Basu, P., and MacDonald, S. (2007) 'Introduction: Experiments in exhibition, ethnography, art, and science', in S. Macdonald and P. Basu (eds) *Exhibition Experiments*, Blackwell, pp 1–24.

Bencard, A. (2019) 'Exhibitions as philosophical carpentry: On object-oriented exhibition-making', in P. Bjerregaard (ed) *Exhibitions as Research: Experimental Methods in Museums*, Routledge, pp 67–79.

Bencard, A., and Whiteley, L. (2024) 'Pill Machine: The Doctor will see you now', in M. Jacobsen (ed) *Apparatus*, m&m paper.

Bencard, A., and Whiteley, L. (2026) 'Curating health complexity at Medical Museion: The exhibition as medium and process for sharing an STS sensibility', in H.S. Rogers (ed) *Curatorial Cultures/What Curators Know: Art and Science in the Museum*, Rowman Littlefield.

Bencard, A., Whiteley, L., and Thon, C.H. (2019) 'Curating experimental entanglements', in M.V. Hansen, A.F. Henningsen, and A. Gregersen (eds) *Curatorial Challenges: Interdisciplinary Perspectives on Contemporary Curating*, Routledge, pp 133–46.

Bencard, A., Bjerregaard, M., and Lillemose, J. (eds) (2022) *The World Is in You: Documentation Reflections Lessons*, Medical Museion.

Berger, B. (2004) *The Survival of a Counterculture: Ideological Work and Everyday Life among Rural Communards*, Routledge.

Bjerregaard, P. (2020) (ed) *Exhibitions as Research: Experimental Methods in Museums*, Routledge.

Boal, A., and McBride, C.A. (2014) 'Theatre of the oppressed', in R. Caines and A. Heble (eds) *The Improvisation Studies Reader: Spontaneous Acts*, Routledge, pp 79–86.

Bogost, I. (2012) *Alien Phenomenology, or, What It's Like to Be a Thing*, University of Minnesota Press.

Born, G., and Barry, A. (2010) 'Art-science', *Journal of Cultural Economy*, 3(1): 103–19.

Brown, V.A., Grootjans, J., Ritchie, J., Townsend, M., and Verrinder, G. (eds) (2013) *Sustainability and Health: Supporting Global Ecological Integrity in Public Health*, Earthscan.

Callard, F., and Fitzgerald, D. (2015) *Rethinking Interdisciplinarity across the Social Sciences and Neurosciences*, Palgrave Macmillan.

Canfield, K.N., Menezes, S., Matsuda, S.B., Moore, A., Mosley Austin, A.N., Dewsbury, B.M. et al (2020) 'Science communication demands a critical approach that centers inclusion, equity, and intersectionality', *Frontiers in Communication*, 5(2).

Daniels, I. (2019) *What Are Exhibitions for? An Anthropological Approach*, Bloomsbury Academic.

Damsgaard, J.B., and Phoenix, A. (2022) 'World of change: Reflections within an educational and health care perspective in a time with COVID-19', *International Journal of Social Psychiatry*, 68(1): 177–82.

Davies, S.R. (2017) *Hackerspaces: Making the Maker Movement*, Polity.

Davies, S.R., and Horst, M. (2016) *Science Communication: Culture, Identity and Citizenship*, Palgrave Macmillan.

Dudley, S. (2013) 'What, or where, is the (museum) object? Colonial encounters in displayed worlds of things', in S. Macdonald and H.R. Leahy (eds) *The International Handbooks of Museum Studies, Volume 1: Theory*, Wiley.

Dunne, A., and Raby, F. (2013) *Speculative Everything: Design, Fiction, and Social Dreaming*, MIT Press.

Evans, H.J., Nicolaisen, L., Tougaard, S., and Achiam, M. (2020) 'Perspective: Museums beyond neutrality', *Nordisk Museologi*, 29(2): 19.

Fitzgerald, D., and Callard, F. (2015) 'Social science and neuroscience beyond interdisciplinarity: Experimental entanglements', *Theory, Culture & Society*, 32(1): 3–32.

Formosinho, J., Bencard, A., and Whiteley, L. (2022) 'Environmentality in medicine: Microbiome research and the perspectival body', *Studies in the History and Philosophy of Science*, 91: 148–58.

Gerber, N., Templeton, E., Chilton, G., Liebman, Marcia C., Manders, E., and Shim, M. (2012) 'Art-based research as a pedagogical approach to studying intersubjectivity in the creative arts therapies', *Journal of Applied Arts & Health*, 3(1): 39–48.

Grudin, R. (1996) *On Dialogue: An Essay in Free Thought*, Houghton Mifflin.

Grünfeld, M., and Tybjerg, K. (2023) 'Collections, knowledge, and time', *Centaurus*, 65(2): 213–34.

Greenhill, E.H. (1992) *Museums and the Shaping of Knowledge*, Routledge.

Hussey, K., Martin, I., and Whiteley, L. (2026) 'Playing seriously: Making "time animals" between art, science and STS', *Configurations*, 34(2): 245–75.

Law, J., and Hassard, J. (eds) (1999) *Actor Network Theory and After*, Blackwell.

Law, J., and Lin, W.-Y. (2022) 'Care-ful research: Sensibilities from science and technology studies (STS)', in U. Flick (ed) *The SAGE Handbook of Qualitative Research Design*, SAGE Publications, pp 127–41.

Madvig, F., Achiam, M., Adler-Nissen, R., Johansen, N., and Whiteley, L. (2022) 'Coming closer to citizens? Frustrated dialogue on the Danish Health Authority's Facebook page during COVID-19', *Frontiers in Science and Environmental Communication*, 7: 822471.

Macdonald, S., and Basu, P. (eds) (2007) *Exhibition Experiments*, Blackwell.
Muller, L., Froggett, L., and Bennett, J. (2020) 'Emergent knowledge in the third space of art-science', *Leonardo*, 53(3): 321–26.
Poole, N. (2014) 'Change', CODE | WORDS: Technology and Theory in the Museum, Available from: https://medium.com/code-words-technology-and-theory-in-the-museum/change-cc3b714ba2a4
Oetzel, J.G., and Ting-Toomey, S. (eds) (2013) *The SAGE Handbook of Conflict Communication: Integrating Theory, Research, and Practice*, Sage Publications.
Rogers, H.S., Halpern, M.K., Hannah, D., and de Ridder-Vignone, K. (2021) (eds) *The Routledge Handbook of Art, Science and Technology Studies*, Routledge.
Rupprecht C.D.D., Vervoort, J., Berthelsen, C., Mangnus, A., Osborne, N., Thompson, K. et al (2020) 'Multispecies sustainability', *Global Sustainability*, 3(e34): 1–12.
Simon, N. (2010) *The Participatory Museum*, np, Available from: https://participatorymuseum.org/
Söderqvist, T., and Pedersen, B.V. (2013) 'Biomedicine on display: Copenhagen's Medical Museion', in S.J.M.M. Alberti and E. Hallam (eds) *Medical Museums: Past, Present, Future*, Royal College of Surgeons of England, pp 144–57.
Sontag, S. (1969) *Against Interpretation*, Dell.
Spangenberg, J.H. (2011) 'Sustainability science: A review, an analysis and some empirical lessons', *Environmental Conservation*, 38(3): 275–87.
Star, S., and Griesemer, J. (1989) 'Institutional ecology, "translations" and boundary objects: Amateurs and professionals in Berkeley's Museum of Vertebrate Zoology, 1907–39', *Social Studies of Science*, 19(3): 387–420.
Traweek, S. (1988) *Beamtimes and Lifetimes: The World of High Energy Physicists*, Harvard University Press.
Verlie, B. (2022) *Learning to Live with Climate Change: From Anxiety to Transformation*, Routledge.
Weibel, P., and Latour, B. (2007) 'Experimenting with representation: Iconoclash and making things public', in S. Macdonald and P. Basu (eds) *Exhibition Experiments*, Blackwell, pp 94–108.
Whiteley, L. (2025) 'Imagining the integrative research museum: Articulating values and valuing disagreement', in J.D. Talasek and B. Stauffer (eds) *Integrative Contemporary Art and Science Practices: Building Catalytic Structures*, Routledge.
Whiteley, L., and Bencard, A. (2020) *Mind the Gut*, Medical Museion.

7

We Need to Pay Attention to the Past: Rethinking Science Communication by Thinking with History

Jacob Thorek Jensen

Climate change is not merely a concern of the future; it is unfolding here and now, as our planetary system is pushed to its limits (Sakschewski et al, 2025), demanding renewed attention to the cumulative consequences of human activity. These conditions call not only for behavioural change or technological solutions (Ejsing, 2024) but for deeper transformations in how we understand and relate to temporality. In many Western societies, an accelerating pace of life combined with an emphasis on short-term gain and profit has narrowed attention to the present, making it increasingly difficult to perceive the long-term patterns and processes that shape the planet over time (Loorbach et al, 2024). Furthermore, the idea of sustainable development is often framed in ways that imply we can simply innovate our way out of the environmental and climate crises. However, we have entered a historical condition in which a point of no return has been reached (Bonneuil and Fressoz, 2017); there is no going back. Therefore, the task is no longer to restore what was, but instead to imagine other possible futures.

Envisioning sustainable life for both humans and non-humans requires new ways of thinking and acting. As anthropologist Marilyn Strathern (1992) reminds us, 'it matters what ideas one uses to think other ideas (with)' (p 10). By paying attention to the past, we can overcome the limitations of envisioning futures solely from the standpoint of the present. The past offers an immense resource for illustrating how things have been done differently from today and thus provides material for thinking about more sustainable futures – futures that move beyond modernity's materialistic-mechanistic worldview (Dewandre, 2018; Holten, 2024). The sustainability crisis of the present, therefore, demands that we critically consider what ideas and concepts we use to think with.

This concern ought to be a key question for science communication as well. The days when science communication could focus primarily on conveying ready-made scientific knowledge to the public are long gone, and so are the days of upholding a distinction between the world of science and the rest of society (see further Bensaude-Vincent, 2001). In order for science communication to take part in shaping more sustainable futures, new approaches, sensibilities, and formats are required. In this chapter, I explore how the past can play a powerful role in this endeavour – not only as an object of analysis but as an active resource for rethinking science communication in times of numerous sustainability crises. By paying attention to the past, we can open up new ways of thinking, not only about how we might live and act otherwise but also how science communication itself might be rethought and developed.

I begin this exploration by examining the role that history and the past have played in the science communication literature. I then develop a theoretical thinking tool for how the past may be approached and mobilized in science communication – an approach I call *thinking with history*, a notion I borrow from the historian Carl Schorske (1998). This is followed by an outline of my methodological considerations for applying this theoretical framework to empirical material. To do so, I draw on historical spaces from approximately 1650 to 1750 – namely public coffeehouses in England – which I use as a case study to explore how the past may open up new imaginaries for science communication practices. I then employ thematic analysis to develop five themes by thinking with history that may inform future directions in science communication: intercultural interactions, ideas and knowledge development, non-mediated conversations, the arts and artistic practices, and thinking with the unfamiliar. Finally, I discuss how a practice of thinking with history may inform future epistemic orientations and practices of science communication as sustainability – the overarching concern of this book.

Since I have already used the terms past and history several times, it is appropriate to clarify how I distinguish between the two. By past, I refer to time and to how, at least from a Western perspective, we perceive ourselves as changing over time, thereby distinguishing between past, present, and future (Simon, 2021). By history, I refer to the scholarly practice or scientific discourse, defined by particular methods and conventions through which the past is investigated and rendered knowable and by the rules that ensure the production of verifiable knowledge (Koselleck, 2004 [1979]).

History and the past within science communication literature

Western, institutionalized science communication is an academic 'discipline-in-the-making' (Bucchi and Trench, 2021, p 2) that emerged as a field only in the 1980s (Guenther and Joubert, 2017). Since then, considerable attention has been devoted to tracing the historical roots of its practices. Engagements with the history of science communication have played a central role in establishing the field's identity, particularly through efforts to situate practices and formats within longer historical trajectories. Numerous calls have thus been made to explore the history of science communication (Greco, 2002; Govoni, 2016; Orthia, 2016; Bucchi and Trench, 2021). Scholars have examined the evolution of science communication formats (Shapin, 1989; Battifoglia, 2004; Knight, 2006; Bucchi, 2008, 2021), changing relationships between science and society (Bensaude-Vincent, 2001; Meyer, 2016), as well as the historical development of science communication research itself (Trench and Bucchi, 2015; Davies and Horst, 2016; Blue, 2019).

Many of these contributions have sought to map the field, often producing varied configurations that reflect its diversity. Such approaches echo the influential historian Leopold von Ranke (1795–1886), who argued that the historian's task was to examine primary sources in order 'to show what actually happened' (von Ranke, 1973, p 57 [1824]). Here, the historian becomes a figure searching for missing pieces of a puzzle, striving to complete an image of the past. I do not want to dismiss the importance of such historical reconstruction; however, this represents only one way of engaging with the past – one that risks limiting history to a backwards-looking exercise.

As I conceive of history, it is far more than the reconstruction of what once was. Rather, it is always an entanglement of pasts, presents, and futures, and as such, it can function as a resource for long-term thinking in science communication. As a trained historian, I hold deep appreciation for the important work of uncovering stories and perspectives from the past. Yet my ambition in this chapter is to explore the transformative role history may play – not merely as a means of explaining where we come from but as a thinking tool for imagining other possible futures. By foregrounding temporal depth, contingency, and alternative ways of practice, history can help cultivate forms of responsibility and thinking that are central to sustainability. If history is to play a greater role within science communication than merely tracing the origins of its formats and structures – perhaps connecting past and present but rarely engaging explicitly with the future – then it is time to reconsider how we approach the past and what we ask it to do. I therefore seek to contribute to ongoing discussions about the role of history in science communication, not only as a means to establish the field's identity but as a set of theoretical approaches that may help shape future approaches and practices in support of more sustainable futures.

Before moving on, I want to acknowledge that many of the endeavours to explore the history of science communication have focused on the Global North. This Western – and more specifically Eurocentric – understanding of science communication is being increasingly challenged by scholars. Such work calls for greater recognition of diverse forms of science communication, including those practised by Indigenous and First Nations peoples (Orthia, 2020; Orthia and Rasekoala, 2020; Rasekoala, 2023). The aim of such work has been described as 'liberating science communication from a white, Western paradigm' (Finlay et al, 2021, p 2). This broader understanding of science communication – one that recognizes multiple ways of communicating scientific knowledge and moves beyond Western epistemologies – is increasingly acknowledged in recent scholarship (for example, Bucchi and Trench, 2025). These works are essential for exposing and challenging entrenched assumptions within the field and for opening up new research agendas, particularly those concerned with bridging historical and epistemic divides between the Global North and South – and thereby contributing to more sustainable futures.

Decentring Western epistemologies, however, does not entail rejecting engagement with Western histories outright but calls for careful, situated, and critically reflexive ways of mobilizing them. As the sociologist Ruha Benjamin (2025) writes, 'leaving things behind, here, isn't about discarding all past traditions and knowledges. So much of mainstream futurist work is about "moving beyond" or "overcoming" the past, a narrative riddled with hostility toward those racialized as nonwhite and non-Western or "primitive"' (p 104). In order to shape more just and sustainable futures, we need more work of this kind that explicitly challenges assumptions of Western superiority. At the same time, it is important to leave open the possibility that certain historically situated practices from Western contexts may, when approached critically and reflexively, still offer resources for imagining alternative futures.

Therefore, to conclude this section, I want to acknowledge that this chapter is written from a situated position. As a middle-aged white man working within a Western academic and cultural context, I approach the past from my own geographical and cultural location. My intention is not to reassert the primacy of Western perspectives but to critically examine how historical practices from this context may still resonate or hold analytical value today, while explicitly recognizing the colonial and extractive harms that Western institutions and actors have inflicted upon people and knowledges in the Global South, as well as parts of the Global North (see Chapter 3 for more on non-Western approaches to science communication). It is from this ethical positioning that I introduce thinking with history as a theoretical approach for engaging the past as a resource for futures thinking in science communication.

Theoretical framework: thinking with history

In this section, I outline an approach to entangling past, present, and future through history and consider how such an approach may inform science communication in the context of sustainability. The historian Alix Green (2012) argues that 'the lesson from history is that there is no lesson; we are on our own' (p 174). This is a deliberately provocative statement. Why, then, should we pay attention to the past at all? What Green suggests is that history does not repeat itself and that we cannot extract direct lessons from the past to guide present action. Yet this does not render history irrelevant. Rather, it calls for a different understanding of what history can do.

Green (2012) argues that 'the past and the future offer us, in essence, the same problem. The past, as the future, is "not here"; neither exists, hence we cannot with any certainty either recover the former or anticipate the future. We can only represent them' (p 176). This perspective challenges the tradition associated with von Ranke and exemplifies a broader movement among historians to rethink the study of the past – not only as a means of understanding what has been, or even what is, but as a way of directing attention towards what might come next. Such reorientations are particularly pertinent to sustainability, where action in the present is inseparable from responsibility towards uncertain and long-term futures.

The historian Zoltán Boldizsár Simon (2021) further illustrates this shift by arguing that we must see 'the past, the present and the future together *as* history' (p 3; emphasis in original). Viewing these temporal dimensions as entangled is not a new idea; rather, it builds on foundational work by scholars such as Hannah Arendt (2006 [1954]) and Reinhart Koselleck (2002, 2004 [1979]). Importantly, this perspective suggests that the past can be mobilized not only to interpret the present but also as a resource for imagining and shaping possible futures – an orientation that resonates strongly with sustainability thinking, which demands attention to long temporal horizons.

In many ways, contemporary society appears increasingly preoccupied with mobilizing the past in relation to present political discourses. Consider political slogans such as 'Make America Great Again', which invoke an imagined golden age of America and activate selective representations of the past to orient visions of the future. In this sense, the past becomes a powerful political resource. This resonates with the writer William Faulkner's (1951) oft-cited remark that 'the past is never dead. It's not even past' (p 92). Such uses of the past, however, remain representations of what once was, aligning with Green's argument. We thus find ourselves caught between those who turn nostalgically towards the past and those who continue to place their faith in the future. Yet these temporal orientations need not be opposed. On the contrary, their connection is crucial if we are to imagine and work towards more sustainable futures.

To address this, I propose an approach of thinking with history. I borrow this term from Schorske (1998), who uses it to describe a practice that 'implies the employment of the materials of the past and the configurations in which we organize and comprehend them to orient ourselves in the living present' (p 3). Schorske (1998) contrasts this with what he calls 'thinking *without* history' (p 5; emphasis in original), describing thinking *with* history as an approach to understanding modernity that resists detachment from the past. In his work, Schorske (1998) challenges modern architecture, music, and science in the second half of the 20th century for having 'defined themselves not so much *out* of the past, indeed scarcely *against* the past, but detached from it in a new, autonomous cultural space' (p 4; emphasis in original). He argues that thinking with history enables navigation of the complexities of the present by drawing on historical materials and perspectives.

I find inspiration in Schorske's notion and in the ambition underlying his work. However, whereas Schorske primarily mobilizes history to orient understanding of the present, I extend this approach by foregrounding the capacity of thinking with history as a practice of futures thinking. In the context of sustainability, this means approaching the past not only as a repository of examples or precedents but as a generative resource that can cultivate long-term thinking and care for both human and more-than-human futures. When mobilized in this way, thinking with history becomes not merely a means of interpretation but a theoretical approach for reimagining science communication as a practice oriented towards more sustainable futures.

This approach of thinking with history suggests a particular way of engaging with the past, yet it also raises a further question: what role, more precisely, does the past play in this engagement? To address this, I propose viewing the past as a stranger, drawing inspiration from sociologist Georg Simmel's concept of the stranger. Simmel (1950 [1908]) defines the stranger as someone 'who comes today and stays tomorrow', and that 'his position in this group is determined, essentially, by the fact that he has not belonged to it from the beginning, that he imports qualities into it, which do not and cannot stem from the group itself' (p 402).

Approaching the past as a stranger foregrounds historical difference rather than continuity. Rather than treating history as a repository of lessons or precedents, this perspective positions the past as a source of perspectives that remain partially unfamiliar and irreducible to present concerns. In this sense, the past can introduce ways of thinking, valuing, and relating that are difficult to generate from within contemporary frames alone. This resonates with Leslie Poles Hartley's (1953) observation that 'the past is a foreign country: they do things differently there' (p 9), underscoring the productive estrangement that historical distance can offer.

For Simmel (1950 [1908]), the stranger often takes on the role of a trader, bringing in 'products that originate outside the group' (p 403). Thinking with history, I suggest, involves a similar movement: the past brings conceptual and imaginative resources into the present without being fully assimilated to it. Treating the past as a stranger thus becomes an epistemic stance – one that resists nostalgia and instrumentalization while remaining open to difference and possibilities.

As part of this approach of thinking with history, viewing the past as a stranger allows science communication to engage temporality in ways that expand the imaginable. Rather than stabilizing dominant narratives or projecting linear futures, this approach invites encounters with historical alterity that can unsettle present assumptions and open up alternative futures. In this way, thinking with history contributes to reimagining science communication as a practice of enacting sustainability – one that cultivates long-term responsibility, care, and ethical actions.

Method

To investigate thinking with history as an approach to rethink science communication, this chapter adopts a case-study method (Flyvbjerg, 2006; Schwandt and Gates, 2018). In examining coffeehouses, I draw on existing published historical scholarship. I do not venture into archives to uncover new documents that can take part in completing the puzzle of coffeehouses, which would be the traditional method of a historian. Rather than reconstructing these spaces as they *actually were*, I focus on comparing coffeehouses across different temporal contexts to explore what characterized these spaces in their own time and how they functioned as spaces of science communication. This approach reflects an understanding of history not as a source of lessons to be extracted but as a resource for orienting thinking in the present and towards the future (Green, 2012; Simon, 2021). Coffeehouses are thus treated as historical exemplars – configurations that allow reflection on how particular spatial, social, and communicative arrangements enabled specific forms of interaction, debate, and knowledge development. In this chapter, I focus on English coffeehouses not because they were more important than others but because they are particularly well documented and extensively studied. This rich historiography thus enables a more detailed analytical engagement with their spatial, social, and communicative characteristics.

Analytically, I employ thematic analysis (Braun and Clarke, 2006, 2022) to examine descriptions of coffeehouses, focusing on the nature of interactions and events that unfolded within them. This analysis is guided by the theoretical framework, thinking with history, rather than by an inductive search for patterns in empirical data. The aim of this analysis is not to establish direct transferability. Instead, it seeks to achieve resonance (Leavy, 2020) – a form of analytical attunement that encompasses emotional, bodily, and cognitive dimensions. The sociologist Hartmut Rosa (2019) describes resonance as 'a dynamic event, the expression of a vibrant responsive relationship that can be seen perhaps most splendidly when a person's eyes light up' (p 196). Rather than proposing models to be replicated, the analysis thus involves a process of analytical translation, in which characteristics of coffeehouses are reinterpreted in relation to present sustainability challenges and future directions for science communication. In this way, thinking with history becomes a method for opening up alternative ways of conceptualizing science communication as sustainability. The themes presented, therefore, reflect my own engagement with these historical spaces as thinking tools.

Coffeehouses: spaces for conversations about futures

By the mid-17th century, Europe saw the arrival of coffee, which led to the emergence of a new social arena, the coffeehouse (Cowan, 2005). The coffee was imported to Europe from the Ottoman Empire (Hattox, 1985), where it first found its way into Europe through merchants who brought it to Venice. Here, the first coffeehouse opened in 1645 (Melton, 2001). The coffeehouse was established as a new space to enjoy this new commodity, which differed from other spaces for public drinking, such as taverns, inns, and public houses where alcoholic beverages were served (Cowan, 2004b).

In the coffeehouse, citizens from diverse backgrounds congregated to exchange news and gossip and delved into the realms of ground-breaking scientific discoveries, the latest artistic creations and pressing political matters (Cowan, 2005). Coffeehouses thereby became spaces for citizens to enjoy a drink in a public social setting, forge new connections, and stay abreast of current events – all without contending with the rowdiness and dubious ambience often associated with taverns (Pincus, 1995; Johns, 2006; Standage, 2013). Coffeehouses became especially popular in England during the 17th century. However, their presence extended to other countries, where they also saw popularity, such as France, Italy, the Netherlands, and Austria (Johns, 2006; Crang and Thrift, 2010; Rittner et al, 2013).

Drinking coffee was associated with the lifestyle of citizens from the higher social ranks, such as among the university-affiliated inhabitants in Oxford, where the first coffeehouse in England opened in 1650 (Cowan, 2005). Some 25 years later, one could read in a satirical pamphlet that coffee drinking had become so popular in the city that 'all the neighborhood swarm [to the coffeehouses] like bees, and buzz there like them too' (quoted in Pincus,

1995, pp 811–12). The new commodity quickly spread to London, where it took root and had its height in the city between 1680 and 1720 – the decline caused by the introduction of two other new beverages, namely gin and tea (Melton, 2001). By the end of the 18th century, many of the English coffeehouses ceased to operate as open spaces where everyone could enter, provided they could pay a penny for the coffee. Instead, many of them were transformed into private clubs for members only (Pelzer and Pelzer, 1982).

Inspired by historical sources, we can imagine entering an English coffeehouse in the late 17th century. After crossing the doorstep, the coffee master would greet you by asking, 'What news have you, master,' or proclaiming, 'Here is fresh news from all parts of the world' (Pincus, 1995, pp 819–20). In contrast to the tavern, where you were placed in secluded booths, you would be seated at a long table covered with tobacco pipes, pamphlets, and newspapers. After paying a penny for your coffee, you could acquire the latest news or enter conversations with other customers, such as members of the Royal Society, who frequently visited coffeehouses to discuss the latest scientific investigations and discoveries, or poets, dramatists, and literary critics, who conversed about the latest artistic productions. Moreover, you could engage in conversations about political matters such as affairs happening in parliament and political developments on the continent (Pincus, 1995, pp 819–20). Or as John Houghton (1645–1705), a member of the Royal Academy, wrote at the end of the 17th century: 'Coffee-houses make all sorts of people sociable, the rich and the poor meet together, as also do the learned and unlearned: it improves arts, merchandise, and all other knowledge' (quoted in Pincus, 1995, p 833).

It is important to emphasize that the coffeehouses cannot be considered a homogeneous form. They came in various sizes and developed different characteristics in order to attract different customers. Larger coffeehouses had separate areas or even private rooms for customers who wanted a secluded area. However, the conventional coffeehouse had one room with a single large table or several smaller ones to cater to more customers (Cowan, 2005).

Some coffeehouses had a basic interior, while others were richly decorated, such as Don Saltero's in London, which was described by the visiting German scholar and book-collector Zacharias Conrad von Uffenbach (1683–1734) in the following way:

> It looks more like a museum of art and natural curiosities than a coffee-house. For both standing around the walls and hanging from the ceiling are all manner of exotic beasts, such as crocodiles and turtles, as well as Indian and other unfamiliar costumes and weapons. It is pity that these things, of which many are truly curious, should hang there in the tobacco smoke and become spoilt. (quoted in Cowan, 2005, p 123)

It was not unheard of to be able to experience objects, animals, or even humans that were strangers to this part of the world. In 1675, for instance, you could go to Garraway and, for the price of 3 shillings, see an Asian elephant or other creatures and humans from the East and West Indies (Cowan, 2005).

Some coffeehouses became venues for meetings between citizens with shared interests. The Graecian became a favourite space for members of the Royal Society, such as Sir Isaac Newton (1643–1727). Here, members of the Royal Society continued their discussions in an informal setting after the official meetings (Pelzer and Pelzer, 1982). Also, intellectuals, tradesmen, and clerks quickly became regular customers (Standage, 2013). However, within a decade or two after the establishment of the first coffeehouses, visitors from various backgrounds, both locals and foreigners, entered and exchanged news and ideas (Standage, 2013). The French political philosopher Montesquieu (1689–1755) wrote: 'It is an advantage of the coffeehouse that one can sit the entire day as well as night among people from all classes' (quoted in Melton, 2001, p 247).

It was not uncommon to find 40 to 50 visitors gathered in a coffeehouse occupied with conversations about the latest news (Cowan, 2005). Here, men of science and members of the public engaged in face-to-face conversations about various topics, for instance, related to new scientific discoveries (Shapin, 1989). The coffeehouses were thus spaces where the social division that existed in society was dissolved, as the general rule was that anyone could enter into conversation with whoever they wanted and that it was not considered good manners to ask about social rank (Goodman, 1994). However, the inclusive atmosphere of the coffeehouses had limitations as women were a rare sight in most coffeehouses, even though they were not prohibited from entering – there are even examples of women owning coffeehouses in London and other English cities (Pincus, 1995). Nonetheless, we must consider the coffeehouses as a world dominated by men (Melton, 2001; Cowan, 2013).

Conversations were not the only activity that took place in coffeehouses. As already mentioned, some functioned also as cabinets of curiosities, while others became sites for auctions and trade, reading rooms, and meeting places for clubs and associations (Melton, 2001). Nonetheless, conversations among strangers were a key feature of the coffeehouse, which meant that they became spaces for public political discussions, something that had previously been confined to the inner walls of private homes (Pincus, 1995).

In addition, coffeehouses not only offered a dynamic atmosphere for socializing but also served as a space for intellectual discourse and the exchange of ideas, driving forward the shaping of what is, from a Western perspective, the modern world (Standage, 2013). During the political turmoil in England in the middle and latter half of the 17th century, such as the Glorious Revolution in 1688, coffeehouses became spaces for educating citizens for political discussions (Cowan, 2004a). This role of the coffeehouses continued into the 18th century. In 1743, the German playwright Theodor Johann Quistorp (1722–76) wrote about a visit to a coffeehouse:

> [A] coffeehouse is like a political stock exchange, where the most gallant and wittiest heads of every estate come together. They engage in wide-ranging and edifying talk, issue well-founded judgments on matters concerning the political and the scholarly world, converse sagaciously about the most secret news from all courts and states, and unveil the most hidden truths. (Quoted in Melton, 2001, p 243)

Another notable aspect of coffeehouses is that they became spaces where the bold bitterness of black coffee blended with another dark substance, ink (Johns, 2006). Coffeehouses emerged as hubs for distributing, discussing, and sharing pamphlets and newspapers, making them key centres for acquiring the latest news (Pincus, 1995). Owners of coffeehouses provided newspapers for free, turning the written word into a conversation starter for discussions about current affairs (Goodman, 1994). Thus, coffeehouses became not only spaces for exchanging ideas but also for sharing and discussing the latest news.

How can coffeehouses inspire future directions of science communication?

Coffeehouses functioned as spaces where different kinds of people gathered to converse about a wide range of topics such as current political affairs, scientific discoveries, and cultural developments. In doing so, they became sites not only for exchanging information but also for imagining how society might develop in the future. Approached through a practice of thinking with history, coffeehouses can be understood as strangers: spaces and communicative practices that are neither directly transferable to the present nor fully assimilable to contemporary contexts yet capable of offering perspectives that challenge and expand how science communication can be imagined today.

Drawing on a thematic analysis of 17th- and 18th-century English coffeehouses, and informed by the approach of thinking with history, I present five themes that resonate with contemporary efforts to rethink science communication as sustainability. These themes do not function as lessons from the past but as thinking material – historically grounded perspectives that can inspire new ways of designing science communication formats and spaces oriented towards long-term sustainable societal transformation. The five themes are: intercultural interactions, ideas and knowledge development, non-mediated conversations, the arts and artistic practices, and thinking with the unfamiliar.

The first theme concerns diversity and intercultural interactions. Coffeehouses were spaces where people from different social, cultural, and professional backgrounds encountered one another in face-to-face conversations. Within these settings, ideas, news, and knowledge circulated across social boundaries, forming assemblages that temporarily suspended established hierarchies. While these spaces were far from fully inclusive, they nonetheless enabled forms of social mixing that were uncommon elsewhere at the time. By thinking with history, this theme draws attention to the need for contemporary science communication spaces that can counteract echo chambers and address a societal landscape characterized by pluralization, fragmentation, and polarization – developments that increasingly challenge science communication today (Schäfer and Metag, 2021).

The second theme highlights ideas and knowledge development. Coffeehouses were arenas where scientific discoveries were discussed, questioned, and debated by citizens with diverse backgrounds, including non-academic ones. In this sense, they fostered forms of knowledge production, where ideas were not simply transmitted but collectively shaped through interaction. While such dynamics resonate with contemporary understandings of science-in-the-making (Latour, 1987), similar open-ended spaces for collective meaning making appear increasingly rare today (Bucchi, 2016; Wagenknecht et al, 2021; Kankaria and Chakraborty, 2024). Historically, coffeehouses were entangled with broader societal transformations. Thinking with history invites us to consider how such spaces illuminate the need for contemporary science communication to cultivate environments that support experimentation, embrace uncertainty, and enable the reimagining of practices in response to sustainability challenges affecting both human and more-than-human worlds (see Chapter 6 for further elaboration on collective practices).

A third theme centres on non-mediated conversations. Conversations in coffeehouses were not formally facilitated but emerged organically from the interests, experiences, and concerns of those present. Anyone could, in principle, join a discussion, and topics shifted in response to current events or shared curiosities. While these spaces were largely dominated by men, the underlying ambition was to cultivate an atmosphere of openness and engagement. This resonates with notions of hospitality (Bismarck and Meyer-Krahmer, 2016), understood as creating conditions where participation feels possible and meaningful. In contrast to contemporary communication environments that are increasingly mediated and polarized, thinking with history invites renewed attention to face-to-face encounters as a vital component of science communication (compare Bucchi and Trench, 2021; Roig, 2024). This also implies challenging the understanding that scientists have no use for conversing with the public, as their worlds are fundamentally different (see further Bensaude-Vincent, 2001).

The fourth theme concerns the arts and artistic practices. Coffeehouses were both sites for discussing art and venues for artistic performances. As such, they functioned as spaces where different forms of knowledge – scientific, artistic, experiential – intersected and interacted. Through artistic practices, coffeehouses enabled alternative ways of seeing, feeling, and understanding the world, often challenging dominant epistemic norms. As argued elsewhere in this book (Chapter 2), and in arts-based research more broadly (Eisner, 2008; Barone and Eisner, 2012; McNiff, 2018; Leavy, 2020), the arts can play a crucial role in opening up imaginative and affective dimensions of sustainability. Thinking with history thus encourages the development of science communication formats that integrate artistic practices as a means of exploring alternative futures.

The fifth theme concerns thinking with the unfamiliar. When thinking with history, coffeehouses appear as spaces where people encountered what was unfamiliar or strange – whether in the form of new ideas, historical artefacts, scientific instruments, foreign objects, artistic performances, or people from different cultural and social backgrounds. These encounters with the unfamiliar played a central role in shaping coffeehouses as spaces for imagination and debate, as they disrupted established ways of understanding the world. Coffeehouses thereby highlight how such moments of unfamiliarity prompted curiosity, questioning, and reflection, enabling participants to consider alternative ways of doing and thinking. Thinking with history suggests that these encounters with the unfamiliar were not incidental but constitutive of the coffeehouses' role as spaces where new futures could be imagined. Material elements were particularly important in this regard, as unfamiliar objects served as starting points for conversations and speculation. In the context of contemporary science communication, this theme points to the potential of deliberately introducing unfamiliar perspectives, materials, or temporalities in order to unsettle taken-for-granted assumptions – resonating with Simmel's concept of the stranger, understood as an element that brings in ideas and perspectives that cannot originate from the group itself. Integrating elements from the past that appear unfamiliar today may thus help open up imaginative spaces in which more sustainable ways of living with both human and more-than-human worlds can be explored.

Concluding remarks: Pasts as resources for envisioning possible futures

Today, science communication unfolds across a wide range of spaces and formats, including science centres, museums, botanical gardens, zoos, planetariums, maker spaces, theme parks, libraries, community centres (Ayure and Triana, 2022), science cafes (Rader and Gibbs, 2024), living labs (Merzagora, 2017), and publishing houses (Martin, 2019). As new spaces and formats continue to emerge, there is a growing need to reflect on what kinds of science–society relations these settings enable, sustain, or foreclose. To inform such reflections, this chapter has drawn inspiration from historical English coffeehouses, approached through an orientation of thinking with history and articulated through five interrelated themes.

Much like coffeehouses in the 17th and 18th centuries, where conversation, debate, and the circulation of ideas were central, future directions of science communication may benefit from prioritizing spaces anchored in dialogue, participation, and collective exploration rather than the mere transmission of knowledge. Such spaces must be accessible to diverse citizens and take as their starting point the varied experiences, concerns, and knowledges that people bring with them rather than their formal education or social position. While coffeehouses were far from inclusive by contemporary standards, thinking with history foregrounds their capacity to temporarily transcend social boundaries and to function as spaces where ideas could be questioned, negotiated, and reimagined.

Importantly, thinking with history does not suggest replicating past formats or romanticizing historical spaces. Rather, it positions pasts as a resource for questioning the present and for unsettling taken-for-granted assumptions about how science communication ought to be practised. Introducing strange elements into contemporary science communication spaces – whether historical perspectives, unfamiliar objects, artistic practices, or experimental formats – can help disrupt dominant narratives and open up new points of departure. As illustrated by English coffeehouses, materiality and spatial design can play a crucial role in fostering such encounters, enabling curiosity, wonder, and speculative forms of thinking oriented towards the cultivation of more sustainable and desirable futures.

Implications for science communication as sustainability

We need to pay attention to the past – not to replicate it, but to inspire us to rethink how science communication is performed in the present and the future. Thinking with history invites us to use historical difference as a resource for reflection and imagination. In this way, thinking with history as a practice of futures thinking can enable the creation of more sustainable forms of life for both humans and non-humans.

Rather than offering answers, the following questions are intended as prompts to consider in science communication research and practice:

Rethinking spaces and encounters:

- How can historical spaces such as coffeehouses help us rethink where science communication takes place today, and who feels invited to participate?

- What kinds of spatial, material, or social framings could support dialogue, hospitality, and collective exploration rather than one-way communication?

Knowledge development and participation:

- How might science communication create spaces for knowledge development, where uncertainty, disagreement, and speculation are treated as productive rather than problematic?
- What would it mean to take citizens' lived experiences and diverse knowledge systems as a starting point for communication, rather than as something to be corrected or supplemented?

Encountering difference and the unfamiliar:

- How can perspectives from the past – experienced as strange or unfamiliar in the present – be mobilized to unsettle dominant narratives of progress, efficiency, or technological solutionism?
- How might encounters with unfamiliar objects, practices, or ideas help open up imaginative space for thinking otherwise about sustainability?

References

Arendt, H. (2006) *Between Past and Future: Eight Exercises in Political Thought*, Penguin Books.

Ayure, M., and Triana, R. (2022) 'Participatory science communication for transformation in Colombia', *Journal of Science Communication*, 21(2): Y03.

Barone, T., and Eisner, E.W. (2012) *Arts Based Research*, Sage.

Battifoglia, E. (2004) 'Popularised science communication modes in Italian popular science magazines (1788–2002)', *Journal of Science Communication*, 3(1): A01.

Benjamin, R. (2025) *Imagination: A Manifesto*, W.W. Norton & Company.

Bensaude-Vincent, B. (2001) 'A genealogy of the increasing gap between science and the public', *Public Understanding of Science*, 10(1): 99–113.

Blue, G. (2019) 'Science communication is culture: Foregrounding ritual in the public communication of science', *Science Communication*, 41(2): 243–53.

Bonneuil, C., and Fressoz, J.-B. (2017) *The Shock of the Anthropocene: The Earth, History and Us*, Verso.

Braun, V., and Clarke, V. (2006) 'Using thematic analysis in psychology', *Qualitative Research in Psychology*, 3(2): 77–101.

Braun, V., and Clarke, V. (2022) *Thematic Analysis: A Practical Guide*, SAGE.

Bucchi, M. (2008) 'Of deficits, deviations and dialogues: Theories of public communication of science', in M. Bucchi and B. Trench (eds) *Handbook of Public Communication of Science and Technology*, Routledge, pp 57–76.

Bucchi, M. (2016) 'Editorial', *Public Understanding of Science*, 25(3): 264–8.

Bucchi, M. (2021) *Newton's Chicken: Science in the Kitchen*, World Scientific.

Bucchi, M., and Trench, B. (2021) 'Rethinking science communication as the social conversation around science', *Journal of Science Communication*, 20(3): Y01.

Bucchi, M., and Trench, B. (2025) *Science Communication: The Basics* (1st edn), Routledge.

Cowan, B. (2004a) 'Mr. Spectator and the coffeehouse public sphere', *Eighteenth-Century Studies*, 37(3): 345–66.

Cowan, B. (2004b) 'The rise of the coffeehouse reconsidered', *The Historical Journal*, 47(1): 21–46.

Cowan, B. (2005) *The Social Life of Coffee: The Emergence of the British Coffeehouse*, Yale University Press.

Cowan, B. (2013) 'English coffeehouses and French salons: Rethinking Habermas, gender and sociability in early modern French and British histography', in A. Vanhaelen and J.P. Ward (eds) *Making Space Public in Early Modern Europe: Performance, Geography, Privacy*, Routledge, pp 41–53.

Crang, M., and Thrift, N.J. (2010) 'Introduction', in M. Crang and N.J. Thrift (eds) *Thinking Space*, Routledge, pp 1–30.

Davies, S.R., and Horst, M. (2016) *Science Communication. Culture, Identity and Citizenship*, Palgrave Macmillan.

Dewandre, N. (2018) 'Political agents as relational selves: rethinking EU politics and policy-making with Hannah Arendt', *Philosophy Today*, 62(2): 493–519.

Eisner, E. (2008) 'Art and knowledge', in J.G. Knowles and A.L. Cole (eds) *Handbook of the Arts in Qualitative Research: Perspectives, Methodologies, Examples, and Issues*, SAGE, pp 3–12.

Ejsing, M. (2024) *Verden er ikke længere den samme: økologiske krisefortællinger i det antropocæne* [The world is no longer the same: Ecological crisis narratives in the Anthropocene] (1st edn), Hans Reitzel.

Faulkner, W. (1951) *Requiem for a Nun*, Random House.

Finlay, S.M., Raman, S., Rasekoala, E., Mignan, V., Dawson, E., Neeley, L. et al (2021) 'From the margins to the mainstream: Deconstructing science communication as a white, Western paradigm', *Journal of Science Communication*, 20(1): C02.

Flyvbjerg, B. (2006) 'Five misunderstandings about case-study research', *Qualitative Inquiry*, 12(2): 219–45.

Goodman, D. (1994) *The Republic of Letters: A Cultural History of the French Enlightenment*, Cornell University Press.

Govoni, P. (2016) 'The historiography of science popularization', in F. Papanelopoulou, A. Nieto-Galan, and E. Perdiguero (eds) *Popularizing Science and Technology in the European Periphery, 1800–2000*, Routledge, pp 21–42.

Greco, P. (2002) 'Communicating in the post-academic era of science', *Journal of Science Communication*, 1(1): E.

Green, A. (2012) 'Continuity, contingency and context: Bringing the historian's cognitive toolkit into university futures and public policy development', *Futures*, 44(2): 174–80.

Guenther, L., and Joubert, M. (2017) 'Science communication as a field of research: Identifying trends, challenges and gaps by analysing research papers', *Journal of Science Communication*, 16(2): A02.

Hartley, L.P. (1953) *The Go-Between*, Hamish Hamilton and the Book Society.

Hattox, R.S. (1985) *Coffee and Coffeehouses: The Origins of a Social Beverage in the Medieval Near East*, University of Washington Press.

Holten, E. (2024) *Underskud: Om værdien af omsorg* [Deficit: On the value of care] (1st edn), Politikens Forlag.

Johns, A. (2006) 'Coffeehouses and print shops', in K. Park and L. Daston (eds) *The Cambridge History of Science*, Cambridge University Press (The Cambridge History of Science), pp 320–40.

Kankaria, S., and Chakraborty, A. (2024) 'Scientific temper: towards an alternate model of science–society relationships', *Journal of Science Communication*, 23(4): Y03.

Knight, D.M. (2006) *Public Understanding of Science: A History of Communicating Scientific Ideas*, Routledge.

Koselleck, R. (2002) *The Practice of Conceptual History: Timing History, Spacing Concepts* (T.S. Presner, trans), Stanford University Press.

Koselleck, R. (ed) (2004) *Futures Past: On the Semantics of Historical Time*, Columbia University Press.

Latour, B. (1987) *Science in Action: How to Follow Scientists and Engineers through Society*, Harvard University Press.

Leavy, P. (2020) *Method Meets Art: Arts-Based Research Practice* (3rd edn), Guilford Press.

Loorbach, D., Patteeuw, V., Szacka, L., and Veenstra P. (2024) *It's about Time: The Architecture of Climate Change*, International Architecture Biennale Rotterdam, nai010 publishers.

Martin, D. (2019) 'EPCOT theme park as a science communication space: The Test Track case', *Journal of Science Communication*, 18(4): A09.

McNiff, S. (2018) 'Philosophical and practical foundations of artistic inquiry creating paradigms, methods, and presentations based in art', in P. Leavy (ed) *Handbook of Arts-Based Research*, Guilford Press, pp 22–36.

Melton, J.V.H. (2001) *The Rise of the Public in Enlightenment Europe*, Cambridge University Press.

Merzagora, M. (2017) 'Science centres and science engagement activities as research facilities: Blurring the frontiers between knowledge production and knowledge sharing', *Journal of Science Communication*, 16(2): E.

Meyer, G. (2016) 'In science communication, why does the idea of a public deficit always return?', *Public Understanding of Science*, 25(4): 433–46.

Orthia, L. (2016) 'Democratizing science in the eighteenth century: Resonances between Condorcet's Sketch (1795) and twenty-first-century science communication', *Journal of Science Communication*, 15(4): A04.

Orthia, L. (2020) 'Strategies for including communication of non-Western and indigenous knowledges in science communication histories', *Journal of Science Communication*, 19(2): A02.

Orthia, L., and Rasekoala, E. (2020) 'Anti-racist science communication starts with recognising its globally diverse historical footprint', Impact of Social Sciences Blog, Available from: https://eprints.lse.ac.uk/105994/

Pelzer, J., and Pelzer, L. (1982) 'The coffee houses of Augustan London', *History Today*, 32(10): 40–8.

Pincus, S. (1995) '"Coffee politicians does create": coffeehouses and restoration political culture', *The Journal of Modern History*, 67(4): 807–34.

Rader, K.A., and Gibbs, C. (2024) 'Broadening adult engagement and education in science cafés: Lessons from an STS – science communication boundary spanning experiment', *Journal of Science Communication*, 23(2): Y02.

Rasekoala, E. (ed) (2023) *Race and Sociocultural Inclusion in Science Communication: Innovation, Decolonisation, and Transformation*, Bristol University Press.

Rittner, L., Haine, W.S., and Jackson, J.H. (2013) *The Thinking Space: The Café as a Cultural Institution in Paris, Italy and Vienna*, Ashgate.

Roig, A. (2024) 'Our future stories: Approaches to collective storytelling for climate futures and action through an integrative review', *Environmental Communication*, 19(2): 337–50.

Rosa, H. (2019) *Resonance: A Sociology of the Relationship to the World* (J.C. Wagner, trans), Polity.

Sakschewski, B., Caesar, L., Andersen, L., Bechthold, M., Bergfeld, L., Beusen, A., et al (2025) 'Planetary health check 2025: A scientific assessment of the state of the planet, planetary boundaries science', Potsdam Institute for Climate Impact Research, Available from: https://doi.org/10.48485/PIK.2025.017

Schäfer, M.S., and Metag, J. (2021) 'Audiences of science communication between pluralisation, fragmentation and polarisation', in B. Trench and M. Bucchi (eds) *Routledge Handbook of Public Communication of Science and Technology* (3rd edn), Routledge, pp 291–304.

Schorske, C.E. (1998) *Thinking with History: Explorations in the Passage to Modernism*, Princeton University Press.

Schwandt, T., and Gates, E. (2018) 'Case study methodology', in N.K. Denzin and Y.S. Lincoln (eds) *The SAGE Handbook of Qualitative Research* (5th edn), SAGE, pp 341–58.

Shapin, S. (1989) 'Science and the public', in G.N. Cantor, J.R.R. Christie, M.J.S. Hodge, and R.C. Olby (eds) *Companion to the History of Modern Science*, Routledge, pp 990–1007.
Simmel, G. (1950) 'The stranger', in *The Sociology of Georg Simmel* (K. Wolff, trans), Free Press, pp 402–8.
Simon, Z.B. (2021) *History in Times of Unprecedented Change: A Theory for the 21st Century*, Bloomsbury Academic.
Standage, T. (2013) *Writing on the Wall: Social Media – the First 2,000 Years*, Bloomsbury.
Strathern, M. (1992) *Reproducing the Future: Essays on Anthropology, Kinship and the New Reproductive Technologies*, Manchester University Press.
Trench, B., and Bucchi, M. (2015) 'Science communication research over 50 years: Patterns and trends', in B. Schiele, J. Le Marec, and P. Baranger (eds) *Science Communication Today – 2015: Current strategies and Means of Action*, PUN-Éditions universitaires de Lorraine, pp 15–27.
von Bismarck, B., and Meyer-Krahmer, B. (eds) (2016) *Hospitality: Hosting Relations in Exhibitions*, International Conference 'Hospitality – Hosting Relations in Exhibitions', Sternberg Press (Cultures of the Curatorial).
von Ranke, L. (1973) 'Preface: Histories of the Latin and Germanic nations from 1494–1514', in F. Stern (ed) *The Varieties of History: From Voltaire to the Present*, Vintage Books, pp 54–8.
Wagenknecht, K., Woods, T., Nold, C., Rüfenacht, S., Voigt-Heucke, S., Caplan, A. et al (2021) 'A question of dialogue? Reflections on how citizen science can enhance communication between science and society', *Journal of Science Communication*, 20(3): A13.

8

We Need to Listen to the Messy and Unseen: On the Potential of Sound to Make Ecological Relations

Martin Grünfeld

What does zooplankton sound like, and how can listening to phenomena inaudible to the human ear multiply our senses of the world? In this chapter, I draw on examples from scientific research as well as art-science collaborations to discuss how sound can be used as an arts-based approach to science communication as sustainability. It is my contention that sound can intimately connect listeners to messy and unseen parts of the world and make new ecological relations possible. This is important for the development of science communication as sustainability because a crucial premise for this endeavour is that we need to move sustainability beyond anthropocentrism and explore how to cultivate alternative human–environment relations that acknowledge the entanglement of human and non-human systems. In this chapter, I focus on sound because of its potential to make relations differently in slow and durational ways that intimately connect listener to environment. Sound is not about the consumption of content. Rather, based on personal experience I believe that sound has the potential to transform the way we experience the environment and our place within it.

In the chapter, I explore how sound recording and listening can connect us to unknown modes of existence of other organisms and environments through examples from Arctic whale research from the 1970s and recent artistic research on zooplankton by Danish sound and performance artist Maria Brænder. However, beyond focusing on recordings and artworks, I also consider listening as a durational transformative experience that may evolve into alternative practices for science communication as sustainability. This leads to a broader discussion of how arts-based methods can redistribute the sensible and multiply our senses of the world in ways that go beyond questions of transmission and points towards their transformative potentials. The argument in this chapter thus unfolds on two levels. I argue that we need listening as an alternate sensorial modality to complement the dominance of the visible and discursive in contemporary science communication, but also that we can take sound as a cue to consider more profoundly how we can reinvent science communication in a time of crisis through the multiplication of sensory approaches to the world. Throughout the chapter, I have included links and QR codes to various sound bites to accompany and sustain the argument. Indeed, you may treat this chapter as a tour into the world of sound and an exercise in listening as much as reading and thinking.[1]

Dwell on the sounds you just heard (Figure 8.1) for a moment … Your imagination may be running wild, or perhaps your desire for clarity and meaning may be wanting to pin it down. What did you just hear? Was it an engine, construction noise, weird electronic music, insects? At least that was my experience when I recorded the sound a few summers ago, even though I thought I knew exactly what I was recording having attached a contact microphone to it and standing in front of it. Indeed, it seemed so weird that it could sound like that because what I was trying to record appeared so simple: a small green apple with red spots on a lawn slowly decaying. The apple was recorded in July 2024 in the Swedish countryside in Skåne with roughly 50 metres to the nearest neighbour. So forget about engines, construction work, and electronic music. I used an inexpensive

Figure 8.1: QR code: I invite you to find your best pair of headphones (if you have any), scan the QR code or follow this link https://soundcloud.com/martin-gruenfeld/rotting-apple-part-1, and listen. As you listen, try to abstain from judging whether it is good or bad, just listen.

sound recorder connected to a professional contact microphone designed by the British artist Jez Riley French. I attached the contact microphone to the apple using white marking tape and left the recorder in the grass for minimal interference from my body.

As I stood there for hours recording, I was blown away by the sonic variations and complexities I experienced from this tiny piece of fruit. Over a couple of days, I felt continuously drawn back to the apple to record more until we reached a point where it turned quiet. But while I was listening to it, it was unclear what I was hearing. As I was listening, I was amazed by the way the multifarious sounds coming from the apple changed my perception of it. Indeed, listening to the rotting apple gave me a sense of the unseen and messy elements of the ecology it was part of, of its interconnectedness, its terroir (see Chapter 2 in this volume to delve into terroir as a concept for science communication). A terroir composed of a rotting apple, ants, and other insects eating, perhaps even communicating, vibrations from the ground and the wind, and other more-than-human things manipulated and recorded through my microphone placement and recording technique contributing to the situated ecological relations I experienced through listening to this unassuming apple. Perhaps, then, listening can be a powerful way to cultivate science communication as sustainability?

Of course, focusing on sound is not new. Sound is all around us: bird songs, motor rumble, wind, voices. Sound immerses us and connects us to each other and the world. We navigate in the world not just by bodily movements and visual perception but also through listening. Yet as composer Pauline Oliveros notes, while many audible signals are transmitted from the environment to our ears and brains, they are not given attention – we are not listening to them (Oliveros, 2005, p xxi). Indeed, there is a hidden world of non-human sounds waiting to be explored, and the development of digital technologies is currently bringing us closer to these unseen and unheard worlds (Bakker, 2022). Similar to time-lapse photography and microscopy, audio recording can reveal inaccessible worlds in retrievable detail (Toop, 1996). A hidden world that is increasingly being explored today as scientists and artists populate the world with microphones. As the anthropologist Stefan Helmreich has noted, sound as a method is particularly useful when we are exploring messy phenomena with unclear boundaries, such as ecologies and biodiversity indicators (Helmreich, 2015).

Already in 1889, the first documented field recording of a non-human subject was made by an eight-year-old boy, Ludwig Koch, using his father's wax cylinder recorder to record a bird (a common Shama) (English, 2015). However, browsing through four major journals within science communication – *Journal of Science Communication* (JCOM), *Public Understanding of Science*, *Science Communication*, and *Environmental*

Communication – showcases a surprising silence when it comes to sound in science communication. The only exceptions are studies of radio and podcasts, yet most of these studies focus on audio quality and professionalization (see for example Newman and Schwarz, 2018). Meanwhile sound recording in a broader sense beyond the privileging of speech and human expression in musical form and recorded conversations remains silenced. This is surprising given that science communication can be defined as 'organised actions aiming to communicate scientific knowledge, methodology, processes or practices in settings where non-scientists are a recognised part of the audience' (Davies and Horst, 2016, p 4). Thus, it seems reasonable to assume that the increasing use of sound in scientific knowledge production would also lead to an increasing attention to sound in science communication. Yet sound is left understudied if not almost unstudied within science communication. What I want to explore in this chapter is how we might use sound in science communication, but more than just focusing on sound as another topic, I want to explore its potential to engage publics differently and, in that process, also rethink the role of science communication.

The early days of Arctic whale research and the development of bioacoustics

Let us travel back in time to 1977 and the heated international dissent over the future of local traditions of hunting Bowhead whales in the Arctic and consider how Indigenous knowledge of sound transformed whale research. Central to the controversy was the lack of data both on the number of bowheads that existed in the region and the number of whales killed by the local hunters (Rensberger, 1977). Government scientists monitored the number of whales through sporadic visual censuses: visiting biologists standing on the shore counting whales supplemented by aerial surveys (Bakker, 2022). Based on the visual census, the International Whaling Commission (IWC) concluded that there were only a few thousand bowheads left and that they were endangered (Rensberger, 1977). This led the IWC to ban the hunting of Bowheads in the Arctic region.

Yet the effort to save the Bowheads also threatened the livelihood of the Indigenous population – the Iñupiat. For the Iñupiat, the hunting of Bowhead whales is what keeps their communities together; their whole culture is centred on whales: they not only use whales for food but also for shelter, tools, clothing, and in cultural ceremonies and music (Rensberger, 1977). Understandably, they were furious about the ban on hunting; however, this was not only because of the threat it posed to their culture and livelihood but also because nobody had asked them about their knowledge of the whales and their views (Rensberger, 1977). Through their hunting practices, the Iñupiat had first-hand experiences with an abundance of whales and claimed that whales travelled beneath the ice, thereby contesting the official visual census.

As Bakker interprets it, this was essentially a clash between two worldviews: traditional holistic knowledge and Western reductionist science (Bakker, 2022). On one hand, traditional Western science, reducing the presence of whales in the region to the visible number of whales based on the hypothesis that the whales wouldn't be able to travel under the ice. On the other hand, Arctic hunters had been using wooden paddles under water to hear whales travelling beneath the ice for almost 4,000 years (Toop, 1996). The reductivism of Western science, however, not only dismissed Indigenous knowledges but also suffered from a poverty of the imagination (see Chapter 3 for more on critical imaginations and futurisms). While the IWC refused to listen, the Iñupiat formed their own whaling commission and launched a scientific study of bioacoustics in the Arctic (funded by tax revenues from a recent oil boom). Guided by the local population, scientists conducted a durational study to demonstrate that bioacoustics combined with local knowledge could yield a more accurate count of Bowhead whales. Equipped with hydrophones capable of capturing sounds under water and working for long periods in harsh conditions, they gathered sonic data from Bowheads (Figure 8.2).

Again, dwell for a moment on what you just heard … In the recording, there is a variety of sound: generator noise mixed with a high-pitched sound. The high-pitched sound, however, is coming from Bearded seals. Towards

Figure 8.2: QR code: I now invite you to listen to a historical recording from 1978 of Bowhead whales from Alaska. The recording is part of the amazing sound archive at the New Bedford Whaling Museum. Scan the QR code or follow the link: https://soundcloud.com/martin-gruenfeld/bowhead-whale-sound-1978. Then begin listening.

the end of the recording, you also sense another organism as Bowhead whale sounds enter the mix. It is speculated that we are hearing one, possibly two Bowheads at a depth of 2,000 meters. However, compared to a visual census, the data are challenging and messy. One of the key difficulties researchers faced in their analysis was how to separate the sounds from Bowheads from other sounds of ice, wind, seals, and human-made noises. But eventually the scientists had a breakthrough and were able to identify Bowhead calls under the ice. Their data, however, were received by Western scientists and the IWC with scepticism, but eventually they became convinced that the numbers of Bowhead whales were much larger than what a visual census could document. This led to scientific collaborations between the Iñupiat and researchers in the continuous conservation efforts and the legalization of their hunting practices.

Unfortunately, this is not the end of the story, as marine noise pollution and climate change are now threatening the existence of Bowheads in the Arctic. But if we stay with the role of sound in this example, we realize that by inviting other senses into the scientific enterprise, we may very well learn something new. Indeed, the unseen whales beneath the ice and their sounds changed whaling research, and today the world is increasingly populated by artificial ears monitoring biodiversity and ecological changes. From an epistemological viewpoint, it is easy to see the need for sound as well as other sensory modalities to know the world not just differently but also better.

What about public engagement then? Given that sound plays important roles in science, it seems obvious that science communication should also pay attention to it and consider how to engage the public in sonic methods and processes involved in understanding the world. Yet what I would like to propose is perhaps less obvious. In the rest of the chapter, I want to explore the idea that science communication should not just focus on sound as another topic but explore its potential to engage the public differently and in that process also rethink science communication beyond its discursive and visual bias.

Making zooplankton public: a case study of a contemporary art-science collaboration

While science communication research appears to be silent on the issue of sound except when focusing on podcasts and sound quality, an increasing variety of arts-based formats are currently being deployed when presenting science in public, including performances, music, and visual arts (Bucchi and Trench, 2021). This emphasizes a research–practice gap and an urgent need for science communication researchers to pay attention to the use of sound in (and other arts-based approaches to) science communication. The rise of art-science collaborations calls for science communication researchers to pay attention to how this constellation informs and potentially transforms science communication (for more on this, see Chapter 5). As Bucchi and Trench note, art-science appears as a distinct model of science communication that 'represent[s] a broader, non-prescriptive and non-strategic view of science communication encompassing a variety of language and forms of expressions' to engage visitors in science in the making (Bucchi and Trench, 2021, p 7). Now I want to delve into one example, or case study, in detail, to explore how a focus on sound can work differently in science communication.

Let us travel from early bioacoustic whale research to a contemporary art-science collaboration exploring the life of zooplankton in Loch Etive in Scotland. In November 2022, the Danish sound and performance artist Maria Brænder ventured on a field trip with a group of Scottish marine biologists led by Dr Kim Last to Loch Etive, a loch with a characteristic microclimate similar to the Arctic region. The aim of the field trip was to study the mysteries of zooplankton (*Calanus finmarchicus*) migration and how their circadian rhythms or 'body clocks' shape their movement. Given its role in the ecosystem as a main food source for larger marine species, these 'body clocks' influence the entire food web in the North Atlantic and Arctic oceans, and knowing more about them helps scientists to predict how marine ecosystems will respond to future climate change (Paterson, 2017). While the researchers ventured to Loch Etive to study plankton, light, and metabolic rhythms, Maria travelled along to explore something largely unstudied: the sounds of zooplankton.

As Maria noted in a conversation, while marine biology has been using sound to study larger animals such as fish and whales for a long time, no one had really paid attention to the sounds that swarms of copepods might make (M. Brænder, personal communication, 10 April 2024). While the scientists mainly focused on refining echo location to be able to locate swarms of copepods and smaller groups to bring back to the lab with them, Maria paid attention to their actual sounds and used hydrophones both in the ocean and in the lab. Although she managed to record distinct sounds from copepods in the lab, recording in the ocean was more challenging. Her insistence on sound led the scientists to think differently about sound and zooplankton. Rather than using sound for echo location, Maria's participation made the researchers consider sound and recordings as potential forms of data.

During our conversation about Maria's artistic practice and work with zooplankton, she described how she spends extended hours in uncontrollable environments to capture sounds, which are then transformed to create atmospheres as an opening to encounters with other species but always with her own perspective and embodiment as part of the composition process. In this process, she is thus both recording non-humans and using herself to attune to these organisms through listening and embodied engagement.

Figure 8.3: QR code: Now I would like to invite you to experience a sound recording of Maria's performance *Spectral Chasings* (2023) – a site-specific performance installation situated in the Brønshøj Water Tower, Denmark. The sound excerpt is from a film poem linked to *Spectral Chasings*. The composition is made in collaboration with composer/musician John Lemke. Maria has generously shared this excerpt with us, commenting that live art is always difficult to document. So please don't treat this as the complete artwork but more as a glimpse into her work. Scan the QR code or follow the link https://soundcloud.com/maria-braender/spectral-chasings and listen.

Her work is about more than illustrating or communicating the processes or sounds that she documents. Her work on zooplankton turned into a multi-sensory performance, which took place in 2023 in a water tower in Brønshøj, Copenhagen – *Spectral Chasings* (Figure 8.3).

Crackling sounds of copepods accompanied by reverberated choral and instrumental compositions as we take steps into the depths of the ocean and our own imagination. In the multi-sensory performance of *Spectral Chasings*, the audience were invited to roam in a vast, dimly lit concrete water tower with its unique reverberation. The scale of the tower emulated walking at the bottom of the sea: big, slow bubbles floating in the air and the crackling sounds of copepods filling up the space. As an audience member, you would easily get lost and confused within this space. Perhaps losing sight of the zooplankton as protagonists of the work. Yet it would be a mistake to fall into the trap of trying to assess how the science of copepods is travelling within this arts-based format and interpret how its meaning is contained. We must be careful not to turn to the usual standards for evaluating science communication: transmission and effectiveness. Rather this work is more about affect and transformation. Indeed, as Megan Halpern and Hannah Star Rogers argue in their discussion of art-science practice, we can discern at least four types of art-science practice, and only one of them concerns the conveyance of a message (Halpern and Rogers, 2021). More than merely a messenger, art-science can contribute to the generation of knowledge, contextualize knowledge production, and critically assess science. *Spectral Chasings* creates an immediate and affective sense of the scientific exploration of copepods, these tiny creatures in the vast ocean, contextualizing scientific research within a human experiential realm. In this contextualization, sound is a key element. As Maria notes: 'Sound can open the world, it stimulates and provokes you directly, opening up to the unknown' (M. Brænder, personal communication, 10 April 2024). But it depends on whether we are ready to listen.

Listening: from transmission to transformation

When we discuss sound, we easily take listening for granted. But what does listening entail? This question is crucial if we want to develop a sound-based science communication as sustainability and will help us assess the potentials and limitations of this approach. While animals are attentive listeners, receiving you as you enter their environment, humans are often narrowly listening and disconnected from the environments they inhabit (Oliveros, 2005). As Bakker (2022) also notes, compared to other animals, humans are poor listeners. This may be surprising given that listening is 'such a seemingly simple act. Yet how often we fail. How often we listen not to what is there but to what we like to believe is there' (French, 2020). Listening as a conscious attentive act is, then, not necessarily innocent and automatic. Although sounds enter our bodies as vibrations, they are also formed by our attention and expectations. Indeed, as the artist and field recordist Jez Riley French observes, the act of listening can expand our understanding and experience of local environments, but it is too often constrained by the borders we automatically impose (French, 2020). Let us delve a bit more into the concept of listening and then consider how listening can be practised.

The sense of automated and constrained listening is something that the French philosopher Jean-Luc Nancy expresses in his book on listening. Nancy opens his book with a provocative question: 'Is listening something of which philosophy is capable?' (2007, p 1). We could probably pose a similar question to science communication, Science and Technology Studies, and other fields. Behind this provocation lies a distinction between listening and understanding. For Nancy, understanding ('comprendre') is intimately linked to hearing ('entendre'), which entails that hearing is oriented towards language, rather than sound. Nancy's point is that 'hearing' is an automatic process that should not be understood as a physiological process but rather as a cultural-linguistic process of sense making. Meanwhile, for Nancy: 'To be listening is always to be on the edge of meaning, or in an edgy meaning of extremity, and as if the sound were precisely nothing else than this edge, this fringe, this margin' (2007, p 7). The core question for Nancy, and for us considering sound in science communication, is what secrets awaits us if we truly listen, when we try to focus on sonority rather than message: 'What secret is yielded – hence also made public – when we listen to a voice, an instrument, or a sound just for itself? And the other, indissociable aspect will be: What does to be listening, to be all ears, as one would say "to be in the world", mean?' (Nancy, 2007, p 5).

To cultivate an openness towards the world through listening involves a different stance. Instead of asking about the meaning of this or that sound or what message is being communicated, it is about transformative experiences. Experiences that change our relationship to our surroundings and ourselves. In our conversation, Maria described how sound can become confrontational in a way that the visible cannot (personal communication, 10 April 2024). Almost like touching, it is vibrating directly into our ears, potentially creating an intimate experience of connections that both brings us in touch with other beings and makes us vulnerable (M. Brænder, personal communication, 10 April 2024). In contrast to discursive representations of the world, sound can place us directly in the world if we are open to listening. Indeed, as Brandon LaBelle argues, sound can stitch together bodies in a temporal instant, creating a diffuse yet intimate meeting point (LaBelle, 2010). A weird thing about sound, though, is that these meeting points can be established across distances. Recall the Bowhead whale you heard and Maria's zooplankton recordings – these sounds travel from the depths of the ocean to make surprisingly intimate bodily connections with us. Sound vibrations appear capable of crossing distances between people, things, and environments, promising and threatening to shrink or break down such distances (Trower, 2008). To be listening, then, entails to be ready to become vulnerable and open to the other. But how? Let us turn to two sound artists to consider what we can learn from their listening practices.

Let us listen to what Pauline Oliveros has to say about her practice of deep listening: '*Deep* has to do with complexity and boundaries, or edges beyond ordinary or habitual understandings' (2005, p xxiii). As noted previously, we often have a tendency only to hear what we expect to hear; we have a habit of narrowed listening. Deep listening in contrast involves attuning to complexity and edges beyond the usual. This idea of a deeper form of listening resonates nicely with Nancy's description of listening as something that brings us to the edges of meaning and transgresses the usual boundaries that defines our perception. For Oliveros, deep listening involves an expansion of our perception of sound to encounter vastness and complexity. The concept emerged serendipitously for her as she was recording in a big cistern with 45 seconds of reverberation, making it hard to tell the direct sounds from the reflective sounds entangling the past, present, and future (Toop, 1996). However, as Bakker (2022) notes, deep listening is an ancient art still practised as a method to reveal nature's truths.

For Oliveros: '[D]eep listening is a practice that is intended to heighten and expand consciousness of sound in as many dimensions of awareness and attentional dynamics as humanly possible' (2005, p xxiii). In that sense, deep listening can also be seen as related to John Cage's pioneering work in which he tried to develop a music that challenged listeners to attend to all the composed and/or incidental sounds around them (Tsing, 2010). For Oliveros, the practice of deep listening is a form of meditation that cultivates an attentiveness to the space/time continuum that transforms our being in the world. Indeed, she argues that compassion comes from deep listening (Oliveros, 2005). To do this, she has developed a series of exercises and conducted workshops across the globe guided by the premise that listening requires active embodied engagement. I won't go through her extensive catalogue of exercises here but encourage the reader to borrow her book from the local library and test them out for themselves. However, to deepen our forms of listening, time is a key factor.

What happens in the process of an extensive or durational listening experience? Maybe you were patient or curious enough to listen to Richard Hamilton's *Auscultation Points* (Figure 8.4) in its entirety before proceeding to this paragraph (or maybe you are listening while reading ...), or perhaps you have been racing through this text to get to the point. However, if you give in to a durational listening you may experience moving through Morris Arboretum & Gardens, from its bird life, a waterfall, and moss to ultrasonic recordings of bats, a drainage pipe, and electromagnetic recordings of radio waves. Altogether, these sounds communicate the terroir of the park and

Figure 8.4: QR code: To slow down for a moment, I invite you to enjoy a durational listening experience. Scan the QR code or follow the link https://soundcloud.com/rph2389/auscultation-points and listen to Richard Hamilton's *Auscultation Points* – a soundscape of the ecosystem of the Morris Arboretum & Gardens in Philadelphia.

its human and non-human interrelations between animals, plants, water, human-made structures, technologies, and people. For Jez Riley French, time is a key element in his durational artistic practice. His durational method involves being immersed in listening to a single place, object, or situation for an extensive time (hours). Or to revisit these places, objects, and situations again over time: As he notes: 'For me, one essential aspect to my connection to listening, and recording, is duration. I firmly believe one has to allow time for environments and situations to impose themselves on us, rather than the other way round. To let go, to forget all the ideas we arrive with' (French, 2020).

This relation, however, is never one way, as we are also always affecting the environments we are listening in on: 'What we arrive with changes each place, and what comes back with us is, at most, a thin trace of unrepeatable interactions' (French and Law, 2021, p 6). Listening is thus a slow, time-consuming, and potentially site-specific practice that may have the potential to transform our experiences of the environment around us while situating us very directly within it, connecting body and space in time. This, however, may be too ambitious and idealistic in a time haunted by acceleration and information overload.

Challenging listening in the age of acceleration

While it is easy to sympathize with an artistic approach to listening that is deep and durational, we must be careful not to idealize such a practice to a point where it no longer has any bearing on reality. As Max Ritts argues in his book *Resonant Ecology*, the assumption that sound, if properly socialized, will aid the realization of an ecological consciousness is both curious and problematic (Ritts, 2024). Based on studies of sound on the north coast of British Columbia in Canada, Ritts develops a critical sonic materialism that scrutinizes sound as an indeterminate figure in the politics of nature and argues that far from naïvely ascribing to sound the possibility of environmental consciousness and conservation, sound is also being utilized in colonial capitalist projects. For example, digital technologies applied in the monitoring of sounds to protect the coasts and marine life have been appropriated to allow industry to map new shipping lanes, treating sound as an exploitable data resource. Undoubtedly, sound can be capitalized in many ways. As Riley French also notes, listening can impose ourselves onto the environment, enacting a politics of nature through technologies that ensure our ability to feel in control and dominant (French, 2020). We need to move beyond a naïve understanding of sound and ideas of how 'sounds of nature provoke strong emotional and mythic responses in humans' as Toop assumes (1996, p 239). This may be the case, but with the changing technological landscape, we cannot rely on such an assumption.

Moreover, we need to consider what version of 'nature' we are listening to, when we listen. While we may have dreams of a pristine nature experience, in the Anthropocene, we cannot remove ourselves and our influences on the environments we listen to. Rather as we experienced listening to Hamilton's *Auscultation Points*, human and non-humans are deeply entangled in the soundscape. So we need critical forms of listening for the Anthropocene. But we must also remember that despite the exploitation of sound in capitalist projects of growth, sound is also more than digital data, and listening is more than a scientific practice; it can also be a 'form of witnessing that acknowledges our presence as guests on this planet and embraces our kinship with other species' (Bakker, 2022, p 8). Indeed, despite his critical sonic materialism, Ritts (2024) also offers the notion of 'resonant ecologies' to emphasize that we need to attend to the varied ways ideas of sound are connected to material life and economic possibility. Resonant ecologies tell us how sound is an uncertain object that offers us a means for grasping and grappling with interconnections and change (Ritts, 2024). While we cannot assume that ecoacoustics (the study of ecosystems through sound) will automatically engage us in ecological relations, it offers us an opportunity to immerse ourselves and seek out ecological resonances.

This notion of 'resonance' also points us in the direction of contemporary theory and the problematic of acceleration in the work of Hartmut Rosa. For Rosa (Rosa, 2013), 'acceleration' names a defining logic of growth within our societies that entails that they can only stabilize themselves through increasing speed and innovation. Within these societies, we live within an aggressive mode of relating to the world characterized by exploitation, domination, manipulation, and speed. The logic of growth constitutes subjects that run, always busy with new tasks, projects, performing better, and so on. A core concern for a turn towards deep durational listening in science communication is how these running subjects can be made to slow down and listen. However, as Rosa (2019) notes in his analysis of acceleration, the antidote to acceleration is not deceleration, but resonance. This conception aligns nicely with my proposition to explore the potentials of sound in science communication. As Rosa describes it, resonance is a different mode of being in the world characterized by responsiveness and elasticity. An opening that allows yourself to be touched by something. A bodily responsiveness, changing your stance and in the process allowing yourself to be transformed.

Resonance is also a sound term that links nicely to considerations within sound art about natural resonances and how we attune to them through listening. In that sense, listening is a practice that might embody the musical elastic creative stance towards the world that Rosa is looking for. Although Rosa is slightly dismissive of the possibility of creating resonant relations, perhaps sound in science communication could be a way to experiment with what it takes to develop other world relations? An example could be Oliveros's deep listening retreats, changing the minds of people, a few at a time, through embodied deep listening exercises. Perhaps these small steps can guide our unlearning of an aggressive approach to the world, even if only in small, slow steps.

Yet the age of acceleration is also marked by constant growth of information and high-speed circulation through digital technologies. Listeners float in an ocean of information (Toop, 1996). In his book *The Information Bomb*, Paul Virilio describes this ocean as the instantaneous high-speed circulation of information. A 'bomb' that has blasted away the here, only the now exists: as the instantaneous presence of information without distances. A 'bomb' that according to Virilio also led to sensory deprivation, leading to people looking into screens doomscrolling and so on. As he notes with reference to McLuhan's famous dictum: '[I]t is not the *medium* which is the message, but merely *the velocity* of the medium' (Virilio, 2005, p 141). Think of X, TikTok, and other platforms that specialize in short, fast consumable content. Is it then utopian to propose deep durational listening after the information bomb has blasted our senses? Although it may seem appropriate to propose this turn, can it really compete in a transmission-hungry society?

Maybe it can't compete, but it can provide an antidote precisely by awakening our perceptual capacities. Contrary to Virilio's dystopian vision of digital technology, it can do so precisely through technological means: contact mics, headphones, and so on. As I have demonstrated throughout this chapter, sound can easily fade into the high-speed circulatory space of the internet and perhaps undermine the accelerated flow of information. Although this may seem impossible and unthinkable as formats are becoming shorter and the attention span seems to be a thing of the past, we need to work towards a different world. Perhaps we can even find hope in recent online projects such as 'Sounds Right', which frames nature as an artist and makes nature sounds available to generate royalties that support conservation efforts (Sounds Right, nd).

Concluding remarks: Towards a sound science communication

In this chapter, we have travelled through the sounds of oceans, grasslands, organisms, and environments traversing science, art, and theory to consider how a sound science communication might be developed. I hope that it is now becoming obvious that science communication should pay attention to sound and that we should consider how to engage the public in sonic methods and processes involved in our understanding the world. Yet I also want to emphasize the less obvious argument that science communication should be interested in sound as more than just another topic. Sound can have an immediate and intimate impact on us if we are ready to listen. It can be immersive and transformative. In contrast to the visual and linguistic domains, which easily create distances, abstractions, and illusions of control, sound does not stay at a distance but entangles us in the naturecultures we listen to and entails a direct sense of response-ability (Haraway, 2008). In a time of ecological peril, perhaps it would be sounder for science communication to include sound among other means as a way to connect audiences with the environment.

We need listening as an alternate sensorial modality to complement the dominance of the visible and discursive in contemporary science communication, but we also need more than listening to multiply our senses of the world in embodied and affective ways that go beyond questions of effective transmission and points towards transformations. We need to consider how other modes of engagement that we could call arts-based approaches offer us alternative embodied and affective practices. In this context, sound seems to have a huge potential for communicating ecological relations, even if (or especially when) they are complex and messy. Maybe the sounds of a rotting apple in a field, Bowhead whales beneath the ice, and zooplankton in tubes in a lab can fuel our imagination of a different future of co-existence, if we are ready to listen to the messy and unseen.

Implications for science communication as sustainability

Science communication as sustainability is a call to move beyond anthropocentrism to acknowledge the entanglement of human and non-human systems and cultivate other ecological relations. The examples from bioacoustic science and artistic practice explored in this chapter each in their own way thematize ecological relations. More than just being about ecological relations with marine animals big (like the Bowheads) or small (like the zooplankton), however, the examples also do something. The recording and listening practices involved challenge scientific ways of approaching the world by attuning to the messy and unseen. This is something to learn from within science communication, which is often devoted to the efficiency and clarity of the message. In a time of accelerated information, perhaps science communication as sustainability needs to consider the velocity of formats to foster transformative experiences. Listening slowly and deeply for extended periods of time, while itself a sustainable practice, may point us towards different visions of what sustainable living might be like. Indeed, if we begin to listen to something as seemingly odd as a rotting apple, we might discover a path beyond thinking sustainability in all-too-human terms.

The following prompts are intended to point towards a future practice of science communication as sustainability:

- Move beyond the dominance of discursive and visual approaches and enter multi-sensory embodied terrain.
- Try to use other senses and media to make different ecological relations. It makes a difference whether you are looking at – or listening to – the rotting apple.
- Cultivate slow durational experiences (such as listening) as antidotes to the increasingly digital high-speed circulation of information. Your participants might be more open to it than you expect.
- Embrace our shared vulnerability and reconsider how science communication touches people, other organisms, the environment, and not least yourself.

Note

1 I want to acknowledge that according to World Health Organization estimates, 20 per cent of the world population suffers from hearing loss, and for a third of them it is disabling. It is my hope that the argument for a multiplication of sensory approaches to the world within science communication will still resonate even without the experience of the accompanying sounds.

References

Bakker, K.J. (2022) *The Sounds of Life: How Digital Technology is Bringing Us Closer to the Worlds of Animals and Plants*, Princeton University Press.

Brænder, M. (2024) Personal communication, 10 April.

Bucchi, M., and Trench, B. (2021) 'Introduction: Science communication as the social conversation around science', in B. Trench and M. Bucchi (eds) *Routledge Handbook of Public Communication of Science and Technology* (3rd edn), Routledge, pp 1–13.

Davies, S.R., and Horst, M. (2016) *Science Communication: Culture, Identity and Citizenship*, Palgrave Macmillan.

English, L. (2015) 'The sounds around us: An introduction to field recording', The Conversation, 8 February, Available from: https://theconversation.com/the-sounds-around-us-an-introduction-to-field-recording-36494

French, J.R. (2020) 'Migration between nature and listening: Jez Riley French', Land Lines Project, Available from: https://landlinesproject.wordpress.com/migration-between-nature-and-listening-jez-riley-french/

French, J.R., and Law, P.R. (2021) 'ìsland | fjorar: The crystal detector for Cona', Cona, Available from: http://www.steklenik.si/wp-content/uploads/2021/02/FIN_the-crystal-detector-corrected.pdf

Halpern, M.K., and Rogers, H.S. (2021) 'Art–science collaborations, complexities and challenges', in B. Trench and M. Bucchi (eds) *Routledge Handbook of Public Communication of Science and Technology* (3rd edn), Routledge, pp 214–37.

Haraway, D. (2008) *When Species Meet*, University of Minnesota Press.

Helmreich, S. (2015) *Sounding the Limits of Life: Essays in the Anthropology of Biology and Beyond*, Princeton University Press.

LaBelle, B. (2010) *Acoustic Territories: Sound Culture and Everyday Life*, Continuum.

Nancy, J.-L. (2007) *Listening* (1st edn) (C. Mandell, trans), Fordham University Press.

Newman, E.J., and Schwarz, N. (2018) 'Good sound, good research: How audio quality influences perceptions of the research and researcher', *Science Communication*, 40(2): 246–57.

Oliveros, P. (2005) *Deep Listening: A Composer's Sound Practice*, iUniverse.

Paterson, E. (2017) 'Loch Etive research uncovers clockwork copepods', Scottish Association for Marine Research, Available from: https://www.sams.ac.uk/news/archive/archive-2017--sams/sams-news-calanus-finmarchicus-genetic-clock.html

Rensberger, B. (1977) 'Alaskan Eskimos angered over ban on hunting of Bowhead whales', *New York Times*, 5 October, Available from: https://www.nytimes.com/1977/10/05/archives/alaskan-eskimos-angered-over-ban-on-hunting-of-bowhead-whales.html

Ritts, M. (2024) *A Resonant Ecology*, Duke University Press.

Rosa, H. (2013) *Social Acceleration: A New Theory of Modernity* (J. Trejo-Mathys, trans), Columbia University Press.

Rosa, H. (2019) *Resonance: A Sociology of Our Relationship to the World* (J.C. Wagner, trans), Polity.

Sounds Right (nd) 'Museum for the United Nations', Available from: https://www.soundsright.earth

Toop, D. (1996) *Ocean of Sound: Aether Talk, Ambient Sound and Imaginary Worlds*, Serpent's Tail.

Trower, S. (2008) 'Editorial: Vibratory movements', *The Senses and Society*, 3(2): 133–5.

Tsing, A. (2010). 'Arts of inclusion, or how to love a mushroom', *Mānoa*, 22(2): 191–203.

Virilio, P. (2005) *The Information Bomb*, Verso.

9

Implications for Science Communication as Sustainability

Sabrina Vitting-Seerup, Jacob Thorek Jensen, and Marianne Achiam

Whether you have read this book cover to cover or dipped into select chapters, at this point you might be wondering how its insights could shape your own professional practice. Perhaps you are a scientist sharing your work through collaborations, podcasts, public talks, or social media? Or maybe you are a science communicator working in a museum, science centre, library, business, or community organization? Whatever your context, you are likely familiar with the challenge of making complex topics accessible, engaging, and relevant for diverse audiences. In the context of sustainability, this challenge becomes not only more urgent but also much more complex. That is why, in this final chapter, we reflect on the implications of working with science communication as sustainability. The implications we outline are not meant as a toolkit or a list of 'to-dos' but as orientations for practice and ways of navigating the ethical, relational, and political complexities that come with communicating science as part of sustainability work.

As the preceding chapters have demonstrated in various ways, framing science communication *as* sustainability entails a fundamental transformation in what we do and how we do it. While reducing the carbon footprint of humans is important, the road to *how* is often more complex than promoting recycling. After all, if we continue to do the same things – in slightly less harmful ways – we are still headed towards a (slightly less) damaged world. Rather, what sustainability fundamentally calls for is a reimagining of how to be in this world, drawing on cognitive thought, sensibilities, imagination, and sensing. Science communication can play an important role in this shift, but this requires the practices, methodologies and formats that constitute science communication to adapt. Not in the form of modest or minimal systemic improvements, but through a reorientation of how we understand the purpose, scope, and societal role of science communication within the broader context of sustainability. Central to this goal is a critical questioning of established assumptions about the role and meaning of science communication.

This book traces a movement: from rethinking science communication's foundations to feeling, sensing, listening, and making new possibilities for sustainability. It does not close with answers, but opens with invitations to think, imagine, feel, and act differently. These invitations may resonate with you in different ways. Some may feel immediately tangible, ready to be woven into your work, while others may ask for more time and reflection. In the following sections, we try to help this translation into being by offering possible routes for carrying the notion of science communication as sustainability into both scientific endeavours and communication initiatives. For clarity, we present these suggestions under two headings: one focused on research, the other on practice. Yet, research and practice are, in reality, deeply entangled, which is why this division is meant as a heuristic, not a hard boundary.

To support you in translating ideas into methods and concepts into action, we turn specifically to storytelling as a final instance of the arts-based approaches discussed throughout this book. As a fundamentally human mode of understanding, storytelling is rooted in our ability to make sense of experience, recognize narrative coherence, and envision alternative futures (Fisher, 1985; Mattingly, 1991). Even though narrative approaches have sometimes been met with scepticism in scientific contexts (Dahlstrom, 2014), we align with scholars who argue that storytelling complements scientific thinking precisely *because* it embeds knowledge in lived realities and emotional resonances (Cormick, 2019; Raven, 2017; Verlie, 2022; Bloomfield, 2024). Stories do not offer universal truths (what does?) but instead provide situated insights grounded in particular encounters, practices, and dilemmas. The fictional stories that follow thus serve as imaginative entry points or relatable examples, rooted in our ideas yet flexible enough to accommodate your own experiences and reflections (compare Morris et al, 2019). Rather than providing blueprints, these stories aim to open possibilities, helping you explore and envision what science communication as sustainability might look and feel like by offering contexts for asking questions, brainstorming, and real-world problem solving (Renstrom, 2015). Ultimately, storytelling becomes a powerful way to explore possibilities without reducing complexity – to imagine without prescribing – and to demonstrate that science communication as sustainability is not a utopian ideal but a pragmatic, adaptable framework that can take shape across diverse settings.

Implications for research: rethinking how we come to know

Working with science communication as sustainability calls for a reorientation of research agendas. Rather than asking how we can most effectively convey sustainability science, the question becomes: how is sustainability knowledge made, shared, and capable of transforming those who engage with it? Here, sustainability is not a secondary outcome tucked behind claims of objectivity or neutrality – it becomes a direct and explicit goal. Science communication as sustainability draws on co-creative approaches and earthbound practices that engage the senses, emotions, and a plurality of perspectives. These practices do not aim to convey scientific truths in isolation but to co-shape new imagined worlds – worlds that are informed by science but that do not necessarily prioritize it over other ways of knowing.

This shift has a range of methodological and epistemological implications. In what follows, we emphasize three implications we consider especially relevant across the chapters of this volume – touchpoints that may serve as starting places for research shaped by this reimagined approach. First, our reimagining invites a broadening of *what* counts as valid knowledge. Science offers a powerful lens through which to see certain aspects of the world, but at the same time, it can overlook other dimensions that matter deeply in the pursuit of sustainability. This is not a call for relativism but rather a challenge to the idea that quantitative data and the epistemic norms of the natural sciences are the only paths to follow to a sustainable world. Research can – and should – recognize diverse knowledge systems, including embodied, emotional, tacit, and Indigenous ways of knowing (Messerli et al, 2019). Embracing alternative epistemic systems, such as those grounded in situated knowledge and standpoint epistemology (Haraway, 1988; Harding, 1993), does not undermine scientific rigour. On the contrary, it can deepen it by incorporating perspectives from positions often overlooked or marginalized (Vitting-Seerup, 2021). The challenge, then, is to explore how these ways of knowing can be meaningfully engaged – through the senses, through emotions, and through practice – and what role science communication can play in making these entanglements visible, legible, and valued.

Second, science communication research must – in close dialogue with practice – continue to *embrace* methodological experimentation. Arts-based methods, speculative design, narrative enquiry, sensory ethnography, and participatory action research all offer rich ways to engage with the complexity, ambiguity, and tensions that surround (and sometimes hinder) sustainability. Crucially, interdisciplinarity in this sense is not about aligning perspectives around pregiven goals but about cultivating a willingness to be unsettled – to let one's own positions be questioned and transformed through collaboration. These approaches are not simply tools for gathering or disseminating knowledge; they shape how sustainability is understood, felt, and created. This calls for research that is attentive not only to outcomes but also to the processes through which knowledge emerges, and that reflects critically on its own assumptions and positionalities.

We do not have to invent everything from scratch. Borrowing and adapting methods from other fields can open new pathways for enquiry, even though doing so raises broader questions about the structures that govern academic work. If science communication is to enact sustainability through experimental and relational practices, then institutions and funding systems must also evolve to support transdisciplinary, process-oriented, and creative approaches. This includes rethinking reward systems, publication formats, and the boundaries set in place between the roles of researcher and practitioner. As this book has shown, some of the most generative insights emerge precisely in the entangled spaces where disciplines blur, and where research is not separate from practice but embedded within it.

The third, and perhaps most far-reaching point, is that science communication as sustainability calls for *rethinking* the very purpose of science communication. While it will remain important to understand and, at times, influence individual behaviours, science communication as sustainability shifts our field's orientation towards fostering collective and systemic change. This reorientation also challenges how we evaluate science communication and its impact. Instead of focusing primarily on short-term knowledge gains or attitude shifts in individuals, we might ask: does a given initiative create the conditions for dialogue across differences? Does it cultivate new sensibilities? Does it open space for imagining alternative futures?

The kinds of imaginaries we need cannot be conjured by individuals alone. Rather, meaning making unfolds through 'semiotic universes of possibility' (Jovchelovitch and Hawlina, 2018, p 137), that is, rich, collective acts of imagining that no single person can generate in isolation. It is through these shared acts of imagining that new futures can be articulated, rehearsed, and made tangible. Science communication research, then, must consider how its practices contribute to or constrain such collective processes of world making.

Taken together, these ideas imply a significant rethinking of what science communication research is for, how it is done, and with whom it is created. To help crystallize this shift, Table 9.1 offers a comparative (but not final) overview of key dimensions that distinguish the traditional 'science of science communication'[1] from the framework we propose.

Do note that Table 9.1 should be seen as a heuristic rather than a definitive classification. We have created it as a way to visualize the transformation in goals, values, and methods that occurs when we move from transmission to transformation, from clarity to complexity, and from persuasion to co-creation. While this comparative framework hopefully helps outline the conceptual shift we propose, it is just as important to consider how such

Table 9.1: Key dimensions of the science of science communication and of science communication as sustainability

Dimension	Science of science communication	Science communication as sustainability
Goal	Transmission	Transformation
Thinking	Causal logic: Linear/ Cause–effect	Epistemic stance: Polyphonic/ Situated knowledges
Models	Neat/Controlled/Predictable	Messy/Emergent/Relational
Epistemology	Normative/Objectivist	Earthbound/Situated
Strives for	Effectiveness/Behaviour change	Resonance/World making
Fosters	Cognition/Rational understanding	Sensory, emotional, embodied, and cognitive knowing
Mode of engagement	Persuasion	Co-creation and/or attunement
Role of publics	Target audiences/Recipients	Co-creators/Kin
Role of science	Neutral arbiter	Situated actor
Temporal orientation	Focused on the measurable now	Thinking with the past to create alternative futures
Imaginaries	Limited/Progress/Manageable futures	Plural/Uncertain futures/ Imagination as method

changes might unfold in practice. What might this shift look like – and perhaps more importantly – what might it feel like?

The following story traces the journey of a researcher who begins with familiar approaches but gradually embraces a more co-creative, situated, and relational way of doing research. It offers a glimpse into the lived experience of reimagining science communication as sustainability.

Storying for research

Her research had always followed the rules: clear questions, measurable outcomes, carefully designed surveys, and coded observations. It focused on how people understood science – what they knew, trusted, and believed. In many ways, it worked just fine: the findings were publishable, fundable, policy relevant. Still, something felt off ...

At a public climate event, the post-survey results were strong: attendees reported learning what they were supposed to, and trust in science was up. But what lingered could not be found in her data …

She kept thinking of this child who had bent over a tray of moss, and whispered: 'Do you come from the forest behind grandma's house?' It was not just what the child said in the most caring, tiny voice – it was also the look of compassion and curiosity in his eyes, how his body curled as he leaned in over the tray, the gentle way his hand slid over the green fluff … It felt so significant, but there was no box to tick in her research design.

Over time, such moments accumulated. Stories from participants, sensory encounters, tensions between what was documented and coded and what was experienced. How to make room for this in her research? And what if she was not just missing data points but overlooked important connections, perhaps whole sets of beliefs?

A project with a science museum offered an opening: a collaboration between sustainability scientists, local artists, and community members to explore sea-level rise. They all brought knowledge about salinity, erosion, and climate mitigation models, despite calling them by very different names. They all brought memories, feelings of loss, and folklore that seemed to weave in and out of the changes and artefacts they discussed. However, the artists were the only ones to bring boxes of repurposed cardboard, paintbrushes, and home-made glue – which quickly made eyes roll among some of the scientists. She herself was there to study the language of their interactions and help translate the science, but her focus shifted already on the first day.

The artists invited the group to a nearby beach once used for harvesting seaweed. They asked everyone to share a memory or story tied to the sea. A woman spoke of her grandmother's belief that the sea 'takes what it needs and gives when it can'. A scientist began explaining tidal patterns – but the artist gently interrupted: 'Can you maybe draw your points instead of explaining them?' They all started sketching on odd pieces of cardboard in silence. While the group ate lunch, an artist placed the drawings in the sand, creating a shared tapestry: salinity maps overlapping with stories of saved fishermen, erosion lines woven through family histories of shared summerhouses. It all seemed so connected, as if the science was not separate from the stories. It seemed to live among them – perhaps even in them!

The next day, the researcher left behind her observation sheets. Instead, she listened. Made things. Facilitated spaces. It was not that she abandoned her old methods (they were still quite useful), but she allowed herself to focus on what lingered. She began to see her role not as expert-explainer but as co-weaver. Her methods expanded: from surveys to sensory ethnography, from interviews to collaborative making. Her 'lab' became the coast, the participants became co-authors. But where could she publish collages of seaweed, examples of grief, experiences of resonance?

Most journals did not know what to do with her work. Reviewers asked for clearer variables, and the funding body wanted impact metrics. 'How does this scale?' they asked. It did not always scale. But it lingered. It rooted.

Staying in academia meant navigating tensions and structures that were slow to adapt, but cracks appeared: special issues, curious funders, and peer reviewers willing to be surprised. And there were others, too, making similar shifts, experimentally working together with new disciplines. They never abandoned the rigour of research, but they did try to redefine it as they found ways of enacting sustainability instead of simply describing potentials for it.

Implications for practice: rethinking what we do

Doing science communication as sustainability means engaging not only with the ecological crises of our time, such as climate change and biodiversity loss, but also with broader challenges, including social justice, cultural survival, and epistemic diversity. As we argue in this book's Introduction, science communication must reflect the very transformations it seeks to support. In other words, we must *enact* as sustainability.

This shift calls for expanded attention to what we communicate, how we communicate, with whom, and to what end. It means moving away from communicating science as a finished product and towards embracing science-in-the-making – the ongoing, situated, and contested processes through which knowledge is produced, debated, and shared (Latour, 1987). Science communicators are thus invited not only to look backwards (at established facts) or outward (to inform others) but forward: to cultivate collective capacities for imagining alternative futures. Practices such as future thinking (Alford, 2025) and future casting (Shamiyeh, 2025) exemplify this shift, inviting audiences into co-creative acts of world building rather than positioning them as passive recipients of settled truths.

In practical terms, this reorientation brings with it several key implications. First, it requires designing for *emergence* rather than *control*. Instead of assuming a linear path from intended message to learning outcome, science communication as sustainability calls for formats that invite open-ended exploration, such as installations, performances, conversations, or participatory events that allow for emotional engagement and embodied interaction. In such spaces, uncertainty is not a problem to be solved but a condition to be shared and navigated together. Crucially, this involves ensuring meaningful inclusion of diverse participants – diverse across disciplines, lived experiences, cultures, and positionalities. Designing for emergence means creating the conditions for multiple ways of knowing to meet, not by erasing difference but by engaging it as a generative force of collective meaning making.

Second, and perhaps most challengingly, we need to *slow down* to allow for *different perspectives* to rise. This means acknowledging the power asymmetries that often shape science communication (for example, between institutions and publics, or between formal experts and non-formal experts) and actively working to redistribute voice, agency, and recognition. Building trust takes time as genuine inclusion requires that people are not just invited in to listen but empowered to shape the agenda (Vitting-Seerup and Wiegand, 2019). In a culture driven by deadlines, deliverables, and innovation cycles, there is rarely enough time for reflection or deep collaboration. Yet science communication as sustainability often requires a different tempo, one that allows for emergence, responsiveness, and repair. Slowing down, paradoxically, also creates the conditions for acting swiftly when necessary. This includes responding to public discussions on current matters or accepting invitations to contribute to events or presentations on short notice.

Third, our practice must be *earthbound* and *relational*. The metaphor of terroir (see Chapter 2) reminds us that science communication as sustainability cannot be one-size-fits-all. It must take root in particular cultural, temporal, spatial, and ecological contexts, attuned to the needs, values, and relations of specific participants, communities, and more-than-human entities. In many ways, this aligns with what skilled science communicators already know intuitively: the importance of reading the room, adapting to the dynamics of a given space, shifting tone or format when the audience drifts, or reworking a plan when the topic takes a turn. Science communication as sustainability builds on this sensitivity of practitioners, extending it to include not just audience engagement but also broader questions of place, power, and planetary entanglement.

Fourth, we need to consider *senses*, *emotions*, and *imagination* as important as the rationales of science. As explored throughout this book, artistic and sensory forms of communication bring unique modes of knowing into play – modes that are crucial for making sense of the complex, emotionally charged, and uncertain issues of sustainability. These practices can hold ambiguity, surface unspoken meanings, and connect people across differences. But be aware: arts-based approaches are not embellishments or tools for translations (Achiam et al, 2026). Only when used to bring together distinct tools, materials, and perspectives can they expand the reach and resonance of science communication.

Finally, we recognize that practising science communication as sustainability demands *institutional change*. Science communicators working in museums, universities, the private sector, and other organizations need not only time and funding but also cultures and leadership that support experimentation, care, failure, and reflection. This means pushing back against narrow performance metrics and short-term project cycles and advocating for structures of support that allow for long-term relationship building, co-creation, and shared learning.

These five implications range from actions you can begin integrating into your practice today to broader shifts that require structural change. Needless to say, you do not have to tackle them all at once. Taking on one at a time may be what gradually transforms not just your practice but how you think about sustainability. Still, we know it might be tempting to overlook the fifth point. Institutional change often feels beyond reach, especially when you are working in a small team or within the constraints of a short-term project. Just remember: the hardest thing to do is often what is needed most.

So how might these ideas take shape? What does it look like to move from good intentions to embodied change, especially when the terrain feels uncertain or the structures seem immovable? Rather than offering a checklist, the following story invites you into an imagined experience. It traces the journey of a science communicator who begins with familiar approaches and gradually embraces a more co-creative, situated, and relational way of doing her job. It offers a glimpse into what this shift may look and feel like, not only intellectually but methodologically, institutionally, and personally.

Storying for practice

For years, her work had followed a familiar rhythm. Climate facts became infographics. Soil degradation turned into short videos. Rising temperatures became shaded red on local maps. Her role was clear: transmit knowledge, change minds, make a difference. But why was it not working?

The science was clearer than ever. The communication efforts were massive – and highly evaluated. And yet: forests kept falling, oceans kept warming, insects kept vanishing.

Why was it not working?

The turning point came unexpectedly during a residency hosted by a sustainability research centre meant to foster collaboration between scientists and local professionals. She was mostly there to do social media, but what stayed with her was not the whiteboard notes she tweeted or the gorgeous background of the estate – it was a rotting log.

One morning, the group followed an ecologist into the forest. They stopped by a decaying trunk mottled with lichen. 'This is my lab,' the ecologist said, kneeling beside it. 'This is where decomposition teaches us how systems renew themselves.' They all looked in silence. It was a bit awkward, to be honest, as no one really knew what they were looking at.

A local theatre performer then invited everyone to sit. To feel the dampness of the wood. To listen – not just to each other but to the slow work of rot, the hum of microbial life interrupted by the call of a distant bird. A participant lay down near the log and declared, 'I am of the lichen now.' Most laughed, but a handful mimicked his move. Even a local politician, who had made a face of disgust once she saw the decaying wood, was now listening with closed eyes and a hand on the lichen. There was something about that shift from disgust to curiosity that was hard to forget ... Something shifted.

When returning to work, her approach began to change. A project on heat resilience in urban spaces did not begin with a public awareness campaign. Instead, it started with listening: to residents, of course, but also to trees planted decades ago that now offered rare shade; to pigeons nesting under solar panels; to soil sensors capturing moisture changes after a rare summer storm. How could they each play a part? What did they need?

Together with local scientists, artists, and neighbours, she built a pavilion – not just for the public but with them. It had moss walls that cooled the air, bird feeders that offered joy to both feathered and clothed city dwellers, and a QR code that led to a soundscape of the microbial activities beneath the pavement. It was close to a bus stop, and many waiting citizens scanned the code: 'What is that?' they asked with their headphones in. 'Is that happening for real?'

But the funders wanted proof: 'Did satisfaction rise in the area?' they asked. 'How much press did you get?' They always asked that, but this time, it seemed like the wrong questions. What if the pavilion was not the real outcome but the process of creating it was? She would rather evaluate who got to speak, to shape, to matter ... Sharing the findings of sustainability scientists remained central, of course, but the usual dissemination had become something else: a practice of attunement, a mode of co-creation, a sensibility to life in all forms!

The evaluation frameworks asked for 'impact levels'. She tried to comply, translating resonance into reports, squeezing emergent meanings into templates not made for them. Satisfaction rose, yes, for humans, birds AND microbes, she added in the margins.

Sometimes the strange reports were accepted. Often, they were questioned. And there was rarely time to pause, to propose new frameworks as an alternative. Real transformation had to be carved from the tight corners of overfilled calendars.

Still, the shift held.

It was no longer just about closing the gap between science and society, or rather – it was also about that. But her practice had become more. Sustainability was not simply explained ... it was somehow enacted as they transformed one small corner of the collapsing world after another. Together.

A call for entanglement and ongoing transformation

Transformation is already underway. Across research and practice, new approaches are taking shape: tentative, experimental, grounded in care. But transformation does not simply unfold on its own. It must be supported, cultivated, nourished. Science communication as sustainability invites us into this ongoing work, not as observers but as active participants – and active in new ways. Science communication as sustainability asks us not only to be translators of knowledge but co-weavers of futures. It challenges us to rethink what counts as impact, whose voices matter, or who even has a voice, and it asks us what practices we choose to carry forward.

While we have presented implications for research and practice separately, we underline yet again that we only see this division as a heuristic. In reality, these domains are deeply entangled within science communication as sustainability. Research does not merely guide practice, nor is practice just an application of theory. Instead, the two co-emerge, as knowledge is not only studied or shared but collaboratively shaped in relation to place, publics, and planetary concerns. In this entangled mode, research and practice become part of the same cycle of enquiry, reflection, and world making. Knowledge arises through doing, sensing, listening, and building – through the messy, relational work of making-with others. A researcher may begin with a sensory ethnography and discover that the process itself becomes a form of communication. A practitioner may co-create an exhibition and find that it generates new theoretical insights. In these blurred spaces, theory is not simply applied to the world but transformed by it.

Let us be clear: science communication as sustainability does not call for better collaboration between research and practice. Instead, it asks us to rethink the boundary altogether. However, this entanglement also complicates roles: researchers may become conveners, facilitators, or artists; practitioners might turn theorists or epistemic innovators. These entanglements are no longer about agreeing on objectives from each side but about being willing to have your own position unsettled in the process. Though this might bring institutional tensions, it also opens the possibility for a deeper responsiveness to ecological, social, and cultural complexity.

What we have aimed to present throughout this book is not a new orthodoxy. In all honesty, we do not know all that science communication as sustainability may entail. The implications shared in this chapter are thus not absolute answers with a guarantee of success but starting points to be explored. You might say that we offer you this book not as a blueprint but as compost – a collection of ideas, practices, and stories to nourish future experiments. Take with you what resonates, rework what needs reworking, and let the rest decay into fertile ground for what comes next.

Note

1 We understand the 'science of science communication' to be the 'empirical approach to defining and understanding audiences, designing messages, mapping communication landscapes, and – most important – evaluating the effectiveness of communication efforts' (Kahan, 2015, p 1).

References

Achiam, M., Vitting-Seerup, S., Hurley, M., Micallef Grimaud, C., Chanay, R., Eric, D.L. et al (2026) 'Practical rationales for art-science in science communication', *International Journal of Science Education, Part B*, 16(1): 57–70.

Alford, K. (ed) (2025) *Cultivating Futures Thinking in Museums*, Routledge.

Bloomfield, E.F. (2024) *Science v. Story: Narrative Strategies for Science Communicators*, University of California Press.

Cormick, C. (2019) 'Who doesn't love a good story? What neuroscience tells about how we respond to narratives', *Journal of Science Communication*, 18(5): Y01.

Dahlstrom, M.F. (2014) 'Using narratives and storytelling to communicate science with nonexpert audiences', *Proceedings of the National Academy of Sciences*, 111(S4): 13614–20.

Fisher, W.R. (1985) 'The narrative paradigm: In the beginning', *Journal of Communication*, 35(4): 74–89.

Haraway, D. (1988) 'Situated knowledges: The science question in feminism and the privilege of partial perspective', *Feminist Studies*, 14(3): 575.

Harding, S. (1993) 'Rethinking standpoint epistemology: What is strong objectivity?', in L. Alcoff and E. Potter (eds) *Feminist Epistemologies*, Routledge, pp 235–48.

Jovchelovitch, S., and Hawlina, H. (2018) 'Utopias and world-making: Time, transformation and the collective imagination', in C. de Saint-Laurent, S. Obradović, and K. Carriere (eds) *Imagining Collective Futures*, Palgrave Macmillan, pp 153–72.

Kahan, D. (2015) 'What is the "science of science communication"?', *Journal of Science Communication*, 14(3): Y04.

Latour, B. (1987) *Science in Action: How to Follow Scientists and Engineers through Society*, Harvard University Press.

Mattingly, C. (1991) 'Narrative reflections on practical actions: Two learning experiments in reflective storytelling', in D.A. Schön (ed) *The Reflective Turn: Case Studies in and on Educational Practice*, Teachers College Press, pp 235–57.

Messerli, P., Kim, E.M., Lutz, W., Moatti, J.-P., Richardson, K., Saidam, M. et al (2019) 'Expansion of sustainability science needed for the SDGs', *Nature Sustainability*, 2(10): 892–4.

Morris, B.S., Chrysochou, P., Christensen, J.D., Orquin, J.L., Barraza, J., Zak, P.J., and Mitkidis, P. (2019) *Stories vs. Facts: Triggering Emotion and Action-Taking on Climate Change* (Climatic Change 154), Springer, pp 19–36.

Raven, P.G. (2017) 'Telling tomorrows: Science fiction as an energy futures research tool', *Energy Research & Social Science*, 31(2017): 164–9.

Renstrom, J. (2015) Science fiction: Teaching students how to save the world', Talk at TEDxWalthamED, Available from: https://www.youtube.com/watch?v=9Ed8OeZcOBI

Shamiyeh, M. (ed) (2025) *Practices of Futurecasting: Ways of Sharing Imagined Tomorrows*, Birkhäuser.

Verlie, B. (2022) *Learning to Live with Climate Change: From Anxiety to Transformation*, Routledge.

Vitting-Seerup, S. (2021) 'Repræsentationer af mangfoldighed: Kulturinstitutioners arbejde med kulturelle repræsentationer i den postmigrantiske tilstand' [Representations of diversity: How cultural institutions work with cultural representations in the postmigrant condition], [doctoral dissertation], University of Copenhagen.

Vitting-Seerup, S., and Wiegand, F. (2019) 'Organizing postmigration in cultural institutions: Diversity work as intrusion, potential or fact?', in M. Schramm, K.Y. Albrecht, and A.S. Lund (eds) *Reframing Migration, Diversity and the Arts: The Postmigrant Condition*, Routledge, pp 193–223.

Index

References to endnotes show both the page number and the note number (67n1).